FREE-RANGE POULTRY

FREE-RANGE POULTRY

THIRD EDITION

KATIE THEAR

Whittet Books

First published 1990 by Farming Press
Reprinted 1997
This edition published by Whittet Books 2002

Whittet Books Ltd, Hill Farm, Stonham Road, Cotton, Stowmarket, Suffolk IP14 4RQ

A catalogue record for this book is available from the British Library

ISBN 1 873580 59 2

Distributed in North America by
Diamond Farm Enterprises
PO Box 537, Alexandra Bay, NY 13607

Printed in the UK by Bath Press

Contents

Acknowledgements

I am grateful to the following for their invaluable contributions:

Scottish Agricultural Colleges (SAC)
Easton Agricultural College
Intervet Ltd
Onduline Ltd
Freedom Food Ltd
The Soil Association
Grassington Rangers
Poultry World magazine
John Tarren
Gordon Holt
Gardencraft
Domestic Fowl Trust
Forsham Cottage Arks
The Wernlas Collection
The Poultry Club of Great Britain
My husband, David

Some of the material in this book has appeared in various articles in *Country Smallholding* magazine.

Preface

There have been many developments in the world of free-range poultry since this book was first published in 1990. Britain spearheaded the resurgence of free-range egg production and now has more free-range egg flocks, per head of population, than anywhere else in the world. France leads the world in free-range table bird production, and has demonstrated how traditional poultry breeds still have a role to play in the commercial world. Standards in relation to free-range and organic production are now shared throughout the European Union, although the latter are lower than those that were originally stipulated by the Soil Association. However, many specialist producers are choosing to follow the Soil Association recommendations. (The SA certifies more than 70 per cent of organic products in the UK.)

Flock size is still a contentious issue, with some free-range egg producers housing thousands of birds per flock. All the evidence, however, is that chickens do best in small flocks. An interesting development is that planning authorities have become increasingly reluctant to allow large, new houses. This has resulted in the development of lightweight and well insulated mobile housing, often with sophisticated extras such as solar panels for the provision of power and light.

In the USA, the concepts of commercial free-range and organic production are still relatively new, and their standards are lower than those of the EU, but the sector is developing rapidly. When I gave a symposium in Missouri, on the subject of free-range, I was overwhelmed by the number of people who attended, and by the level of interest that was shown.

This new edition of *Free-Range Poultry,* published by Whittet Books, reflects the changes that have taken place since the last edition was published in 1997, including changes in regulations and the latest research in relation to free-range.

There is a growing realisation that poultry and livestock are sentient beings, capable of experiencing pain and stress. Public concern is increasingly focused on the need for humane conditions and a respect for the innate characteristics of living creatures.

Parallel with this is an awareness of the need to conserve traditional breeds of poultry. Organisations on both sides of the Atlantic are now working to this end. Poultry breeders are going back to the genetic reservoir of the old breeds in order to develop layers of speciality eggs or slow growing table birds for the free-range sector, as referred to earlier. It is too frequently forgotten that it was the traditional breeds which gave rise to our present high-yielding hybrids.

The number of people keeping chickens on a small domestic scale has also increased. In terms of opportunity, choice of breeds and availability of housing and equipment supplies, there has never been a better time to keep chickens or to start a free-range enterprise.

This book is a practical and comprehensive guide to the free-range management of chickens, on any scale. It is relevant to the large producer, the small producer and the domestic flock owner. It is also up to date in its coverage, with an emphasis not only on how traditional practices have been adapted for modern conditions, but also on the role of scientific and veterinary developments in the avoidance and control of problems.

I hope that the book will prove useful to all potential and existing free-range poultry keepers and producers, to those catering for their needs, to agricultural students and, indeed, to anyone with an interest in poultry.

KATIE THEAR
Newport, 2002

Introduction

Preference is given to the genuine farm egg, produced by healthy stock enjoying free-range and living under natural conditions.

Herbert Howes, 1939

The quotation above arguably sums up the general view of free-range poultry-keeping. The concept of healthy, contented birds ranging at will over natural vegetation and laying fresh farm eggs is a cosy one. There is also the beguiling implication that consumers will give preference to free-range eggs over those produced in more intensive systems. Overall, it is an appealing picture – but how true is it?

The domestic chicken has evolved from the Red Jungle Fowl of Asia. Her ancestors spent their time in their indigenous habitats scratching about in the ground litter of tropical and subtropical forests, seeking the security of tree cover at night or when danger threatened. Their domestic cousins, our familiar chickens, share similar features, and a free-range lifestyle comes naturally to them. They have little natural oil in the feathers, as do ducks, and their plumage cannot readily shed rainwater.

It would be a mistake, however, to assume that any system of husbandry is 'natural'; there is really no such thing. Where poultry are kept for production or interest in field, orchard or garden, there is inevitably a degree of interference by man. What is true to say is that some systems are less intensive than others. The most intensive is the abhorrent battery system, where hens are caged in a totally artificial environment.

A free-range system is the best way of keeping poultry, but it is also a compromise between respecting their natural instincts and protecting them against inclement weather conditions and predators. They must be adequately housed and fenced, particularly in northern climes where temperatures plummet and where foxes can devastate flocks.

When it comes to consumer choice, there is little doubt that most people opt for the cost factor and buy the cheapest eggs and table birds, and these are battery and broiler-produced. What is also true – and this is the salient factor for free-range producers – is that a growing minority of customers are prepared to pay a premium price for free-range eggs and more naturally raised table birds.

Where chickens are kept on a small scale, it is often because there is a wish to have one's own fresh, quality produce, as well as a general interest in keeping or showing poultry. Here too, free-range methods are often used to manage the flock, but what is meant by free-range? In the next chapter it is appropriate to address this and other key questions relevant to anyone thinking of keeping chickens.

CHAPTER 1 # Key Questions

No question is so difficult to answer as that to which the answer is obvious.

George Bernard Shaw

There are many aspects to be considered before starting to keep free-range chickens, whatever the scale envisaged, whether it be a few birds in the garden or a large commercial unit. It is appropriate therefore to look at the key questions involved. The following information may not be relevant to all situations, so take from it what is useful for you.

What is free-range?

In the past free-range was a general description, indicating only that poultry were allowed to range over the fields. There was no legal restriction on the number of birds to the acre if eggs from these flocks were sold. A flock was usually confined to one field, then moved to another, in sequence, as the grazing became exhausted. It was a familiar sight to see a field brown with Rhode Island Reds or snowed under with White Leghorns, so congested that it was virtually impossible to detect areas of green between them. It was common practice to fold or temporarily house chickens on newly harvested arable fields to glean the spilled seeds and clear the land of residual pests. They frequently followed cattle which were moved on as the larger grasses were eaten, making new, short growth available for the poultry. The birds' scratching activities broke down and dispersed the cow pats, providing a useful harrowing service.

Poultry fitted well in a mixed farming economy, particularly as their foraging helped to control pests such as slugs and leatherjackets, two banes of the grower.

For most of today's small flock owners, free-range means much the same as it always did, although the poultry will now have much more space in the garden, orchard or field. It is only when eggs are sold with the description free-range that the situation changes.

Today, free-range is a specific term. European Union regulations demand that eggs offered for sale as free-range must be from flocks which are kept in the following conditions:

- The hens must have continuous day-time access to open-air runs.
- The ground to which the hens have access must be mainly covered with vegetation.
- The maximum stocking should not exceed 1,000 birds per hectare of ground available to them (400 birds per acre, or 1 bird to every 10 square metres).
- The interior of the building must conform to one of the following standards:

Perchery (*barn*) – where there is a minimum of 15 cm perch space per bird and a maximum stocking density of 25 birds per square metre of space in that part of the building available to the birds. *Deep litter* – where at least one-third of the floor area is covered with litter material such as straw, wood shavings, sand or turf, and a sufficiently large part of the floor area available to the hens is used for the collection of bird droppings. The stocking density with this type of house should not exceed 7 birds per square metre of available floor space.

The regulations ensure that if abuses such as keeping a flock indoors for most of the day and allowing them out briefly in order to use the description free-range do occur, the culprits can be prosecuted.

It is important to emphasise that the regulations refer to the maximum stocking rates, and that these may be inappropriate for certain conditions. For example, if the land is not naturally free-draining it will be necessary to reduce the number of birds per hectare. Organic organisations and the RSPCA's Freedom Food initiative have their own standards which must be adhered to if their marketing descriptions are used.

Are there any restrictions on keeping poultry?

This may seem an obvious point to check, but it is surprising how frequently it is forgotten that there are by-laws which prohibit the keeping of poultry in some localities in or bordering urban areas. Check with the local authority. There is usually no problem as long as there are no complaints about cock crowing, smells or rats.

It is also advisable to check title deeds, tenancy agreements or lease contracts which apply to the property in case there are restrictive covenants relating to poultry. These may not necessarily be still binding, as one lady discovered when she investigated her own deeds.[1] Sure enough, there was a clause which prohibited the keeping of poultry, but when her solicitor checked, he found that it had been inserted by the man who had originally sold the land to the builder in the 1930s. As the chances were that he was no longer alive, and possibilities of

complaints were remote, she went ahead. Another enterprising couple who were about to buy a new house in 1991 found that there was an anti-poultry clause. When they said that they wanted to keep a few chickens and were not prepared to buy the house unless they could, the builder immediately struck out the clause.

Is planning permission required?

Planning permission is not required to house and keep a small domestic flock as long as the house is a small, moveable structure. For larger flocks and houses, the situation is more complex, and anyone considering starting a commercial free-range enterprise with a new house is advised to contact an organisation which offers a consultancy and advisory service.

The key factors to bear in mind are the effect on the landscape, and potential traffic, noise, odour or dust problems. An Environmental Impact Assessment is likely to be necessary under Schedule 2 of the Planning Regulations. Further details are given in Chapter 5.

Is it necessary to register a flock?

If you have a small domestic flock there is no requirement to register it with anyone. However, if you sell any eggs for consumption, you need to be aware that they are a food product that needs careful handling in clean conditions. They should only be sold if they are fresh, undamaged and wholesome.

These eggs may be sold direct to consumers for their own needs at the farm gate (including your own farm shop), in local public markets (not auction sales), or by door-to-door selling. The eggs must be the producer's own eggs, and not be graded into sizes, as indicated on page 108.

Those selling free-range eggs from commercial flocks are required to comply with the EU Egg Marketing Regulations and register with the Regional Egg Marketing Inspector (REMI).

Is it necessary to test birds for salmonella?

It is no longer necessary for laying flocks to be tested for salmonella. Most commercial point-of-lay pullets for egg production come from rearers who have vaccinated their birds against *Salmonella enteriditis* and *S. typhimurium*, forms that can be transmitted from hen to egg. Large breeders of pure breeds will also provide certification, but as small breeders are not covered by the legislation, birds from them will not normally have been tested or vaccinated.

Breeding flocks of 250 birds or more are required by the Poultry Breeding Flocks and Hatcheries Order 1993 to be tested for salmonella on a regular basis. A breeding flock is one kept for the production of chicks or pullets for subsequent sale. See Health.

What is needed to use the description free-range?

If eggs are sold with the description free-range, the flock must be provided with the conditions referred to at the beginning of this chapter. It is also necessary to apply for registration with the Regional Egg Marketing Inspector (REMI) who will then allocate a registration number. The site will be open to inspection to verify that the conditions are being met. (Local sales of ungraded eggs from household flocks are exempt.)

What is necessary to sell graded free-range eggs?

If eggs are to be sold graded by size the producer must either register with REMI as a packer or supply the eggs to a registered packing station. The packer grades and packs eggs in the appropriate containers and sells them. The producers must notify REMI before supplying the packing station. The eggs must come from a flock kept in a way that meets the requirements of the free-range legislation referred to earlier.

Free-range eggs can, of course, be sold ungraded.

What sort of site is suitable?

The ideal is a light, free-draining but fertile soil, capable of producing a healthy sward of short grasses. Any badly drained, boggy land is a haven for parasites such as flukes and coccidia. Hens scratching about in badly drained soils will produce boggy conditions relatively quickly. Chickens also require a sheltered location, rather than an open prairie with lashing winds.

If produce is to be sold direct to the public from the site, it needs to be in an accessible place, not down an isolated track.

On a small scale, chickens do very well in gardens and orchards, where the proximity of trees and buildings provides windbreak protection.

What sort of house is needed?

The type of buildings used will depend upon the scale of operations. There are basically two options – moveable houses and fixed or static ones, the latter tending to be used by large commercial enterprises. Within these two categories, there is a wide choice, including the use of existing and refurbished out-buildings, purpose-built units and houses made of a variety of materials. If produce is being sold with the descriptions such as free-range, Freedom Food or organic, the buildings will need to comply with the relevant standards.

Where the premises are registered for egg

grading and packing, a building that will meet the legal requirements for storage and handling will also be needed.

The small domestic flock owner will normally opt for a moveable timber house with an attached or integral run.

What sort of birds?

If chickens are to be kept for commercial free-range egg production, it is best to choose those that have been specifically bred and reared for good egg production in extensive conditions. These are hybrids or crosses that have been floor-reared and perch trained. Unlike battery birds, they have no problems about perching and foraging.

There are specialist rearers and suppliers who concentrate on supplying pullets for the free-range sector. Suitable brown egg layers include ISA Brown, Lohmann Brown, Hisex Ranger, Hisex Brown, Hy-Line Brown, Shaver 579, Babcock B380, Black Rock, Bovans Nera, Hebden Black, Bovans Goldline, Speckledy and Columbian Blacktail. (The latter is also called the Calder Ranger.) The Bovans White, based on the Leghorn and often called White Star, is a white egg layer suitable for extensive conditions.

If table birds are to be produced, the best choice is one of the slow-growing strains of colour-feathered hybrids that have been bred specifically for free-range. They include Hubbard ISA Redbro (Poulet Bronze) and Sasso 551 (Poulet Gaulois). The white-feathered Cobb and Ross broilers, normally associated with the intensive sector, will adapt to outside conditions, but white-feathered birds that have been developed for free-range include Hybro and Sherwood White.

Hybrids are often the choice for the household flock, too. One of the traditional pure breeds may be preferred, of course, particularly as this contributes to supporting and helping to conserve the old breeds, although they are more expensive to buy and keep. A utility strain that has been bred for production is preferable to a show strain if egg numbers are important. In order to show birds, they must be pure bred and conform to the appropriate standards for the particular breed. Examples are Rhode Island Red, Maran, Welsummer, Light Sussex and Wyandotte, to name but a few. Further details are given in the Breeds chapter.

What are the welfare requirements?

The Department of the Environment, Food and Rural Affairs (DEFRA) has produced a *Code of Recommendations for the Welfare of Domestic Fowls*. Free copies are available from DEFRA Publications.

The Farm Animal Welfare Council (FAWC), which is an independent body, also advises the government on welfare issues and produces regular reports and recommendations.

In addition to these basic requirements there are also those produced by organisations such as the RSPCA and the Soil Association.

What are the Freedom Food standards?

Freedom Food is a wholly owned subsidiary of the RSPCA. It was formed to implement and monitor the RSPCA's approved humane rearing and handling standards, beyond those required by the poultry legislation referred to earlier. Free-range producers may apply for registration and if accepted must abide by the RSPCA's Welfare Standards for Laying Hens which include the five freedoms defined by the Farm Animal Welfare Council in 1992. Small flock owners will also find these relevant, even though their interest is non-commercial.

- Freedom from hunger, thirst and malnutrition
- Freedom from discomfort
- Freedom from disease and injury
- Freedom to perform most normal patterns of behaviour
- Freedom from fear

These requirements are reflected in the kind of housing, type of perch, pop-hole and equipment, as well as in details of management system.

What about organic production?

Organic producers must comply with European Regulation 1804 (1999) which was implemented in 2000. In Britain, there is an additional requirement to follow the national standards of the United Kingdom register of Organic Standards (UKROFS). There are several private organisations that are registered with UKROFS and act as certifying bodies. They include the Soil Association, Organic Farmers & Growers, Demeter/Bio-Dynamic Agricultural Association, Scottish Organic Producers' Association, Organic Food Federation and Irish Organic Farmers & Growers. These bodies may also set their own standards. Those of the Soil Association are the highest.

Free-range producers may apply for registration and inspection with any of these bodies. The requirements cover the type of housing, flock density in relation to land and housing, feeding and management.

Some of the supermarket chains also have their own standards. For example, Waitrose sells eggs from flocks of Columbian Blacktails that are housed at a density of 500 birds per hectare, and will not allow the birds to be beak-trimmed on humanitarian grounds.

Registered producers who abide by the appropriate standards may use the recognised Freedom Food and UKROFS logos.

What organisations exist for the free-range sector?

Formed in 1991, the British Free-Range Egg Producers' Association (BFREPA) is an independent body which represents the interests of commercial, free-range egg producers. It has regional representatives covering England and Wales who offer local contacts for help and advice. There is also a monthly newsletter.

The Poultry Club of Great Britain represents the interests of pure breeds and provides an umbrella organisation for the individual breed clubs that are affiliated to it. Nationally recognised standards are set for each breed. National, regional and local poultry shows come under the auspices of the PCGB.

The Rare Poultry Society concentrates on minority breeds that exist in small numbers only, and are in danger of extinction.

The Rare Breeds Survival Trust is also an organisation that seeks to preserve endangered species. Its Poultry Project is an effort to improve the productive capabilities of the traditional breeds that were once kept for eggs and the table. The Utility Poultry Breeders' Association has similar aims.

What sort of output can be expected from free-range chickens?

The breeding companies issue regular statistics based on hen-housed averages (HHA) for their particular strains of hybrid birds. A hen-housed average is calculated by taking the total number of eggs produced during a season's laying period, and dividing this by the total number of birds in the original flock (including any subsequent mortalities). If, for example, 99 birds from an original flock of 100 produced 28,236 eggs in a 52-week period, HHA is 28,236 divided by 100, which is 282. (Laying period normally starts at around 21 weeks.)

In 1995 a flock of 5,000 free-range hybrid chickens in Lincolnshire[2] recorded 304 eggs per bird at 72 weeks. Also in 1995, a Norfolk Freedom Food free-range farmer[3] reported a 312 HHA at 74 weeks. In 1995, ISA Brown provided the following management guide to their birds on a free-range system:

Table 1.1 ISA Brown free-range results to 72 weeks

Hen housed average	321	310	310
Average egg weight (g)	63.8	63.9	66.1
Egg mass (kg)	20.48	19.81	20.49
Size 0–3(%)	74	75	84
Feed/bird/day (g)	133.5	135.4	138.1
Feed conversion ratio	2.46	2.59	2.60
Liveability (%)	94.6	95.0	94.7

Source: ISA Brown Newsletter, May 1995.

As referred to earlier, pure breeds are far less productive than commercial hybrids and production records for them are generally not available. Rare exceptions are the traditional breeds kept by one utility breeder whose birds recorded the following:

Table 1.2 Utility breeds trap-nested and recorded over three years by Clem Shaw

Breed	Average eggs per year
White Wyandotte	244
Rhode Island Red	277
White Leghorn	257
Black/Blue Leghorn	254
Cuckoo Leghorn	262
Jersey Giant	211

Food consumption of the Wyandottes in the cool part of July 1993 was as follows:

Pen 1 recorded for 39 birds over 18 days to 26 July 1993

Wheat	3.98 oz	(112 g)
Layers' mash	0.79 oz	(24 g)
Total: bird/day	4.77 oz	(136 g)

Pen 2 as above but only 19 birds in pen

Wheat	3.08 oz	(86 g)
Layers' mash	0.79 oz	(24 g)
Total: bird/day	4.10 oz	(110 g)

(This shows that these traditional utility birds consume a large amount of grain.)

Source: Smith Associates, Home Farm, December/January 1994.

As far as table birds are concerned, non-intensive ones are raised more slowly than intensive birds, with the aim of producing a quality product in terms of flavour, texture and freedom from additives. Twelve weeks is the normal cycle for free-range chickens. At this age they will have achieved a weight of 1.8 to 2.25 kg (4 to 5 lb). Intensive broilers will normally achieve this weight in half the time.

Where can one learn about free-range poultry?

Reading as much as possible on the subject provides an excellent theoretical background, although it is advisable to be wary of some advice in older books on free-range production, because the information may be out of date. A book such as this is up to date, but it would be a foolhardy author who claimed that reading can take the place of practical tuition.

ADAS (Agricultural Development and Advisory Service) offers a complete consultancy and advisory service for the large free-range producer, including advice on how to set up and run a unit, and researching and reporting on individual sites and enterprises.

Easton College, Norfolk specialises in poultry training and free-range management courses. It has its own free-range egg production unit. A consultancy service is also available for anyone starting a free-range enterprise.

Scottish Agricultural Colleges (SAC) conduct a considerable amount of research into free-range systems, and also have courses.

Other agricultural colleges and some private organisations will arrange a course if there is a specific demand and the numbers are substantial enough to make it viable. ATB-Landbase has information on courses provided by organisations that are registered with it. These include regional and smallholding training groups. Addresses of organisations are given in the Reference Section at the end of the book.

References

1. *Home Farm*, Issue 96, 1991.
2. *Poultry World*, December 1995.
3. *Poultry World*, March 1995.

CHAPTER 2 # About the Chicken

In the developed countries the chicken has become a specialised machine.

E. Grant Moody, *Raising Small Animals*, 1991

Origin and classification

If we 'place' the domestic chicken in the general order of classification within the animal kingdom, the designation is like this:

Kingdom: Animal
Phyllum: Vertebrate (with spine)
Branch: *Gnathostomata* (with upper and lower jaws)
Class: *Aves* (birds)
Sub-class: *Neornithes* (without teeth)
Sub-division: *Carinatae* (with keeled sternum or breastbone)
Order: *Gallinae* (terrestrial birds, short wings, legs and toes for running and scratching)
Genus: *Gallus* (cock-like birds)
Species: *Gallus domesticus* (breeding under domestic conditions)

Most authorities in the past were of the view that the Red Jungle Fowl of Asia provided many of the ancestral genes for today's chickens, particularly because of the range of colours in its feathers. Some suggested another source which would explain the tall, close-feathered fowl of India, Malaysia and Thailand, which differ in bone structure and musculature. Some have suggested that an unknown ancestor of the present Aseel and Malay might be the link. Temminck, in 1823, postulated this as a hypothetical *Gallus giganteus*. However, recent research, using DNA techniques, has finally established that

the progenitor of the domestic chicken is one particular sub-species of the Red Jungle Fowl of Thailand, *Gallus gallus gallus*.[1] Why there should be such a difference between the present Red Jungle Fowl and the tall Aseel and Malay, which are found in the same area, is a mystery. However, in the absence of further DNA proof, we must accept that other comments in this context are speculation.

History and development

How, and over what period, the early development of chickens took place is unknown, although the DNA research referred to above indicates that domestication took place at least 8,000 years ago. Most of the great civilizations, including China and India, were familiar with them, and in most cases the prime interest was in cock fighting. The Egyptians had a sophisticated system of incubating eggs, although the absence of chickens from their early and detailed records may indicate that the eggs were those of water fowl. They were certainly there by 1350 BC because there is a painting of a cock in Tutankhamun's tomb.[2] The Biblical reference to Peter and the crowing of the cock tells us that chickens were known in Syria and the Holy Land.[3]

The two main routes along which fowl have spread through the world are, firstly, by

9

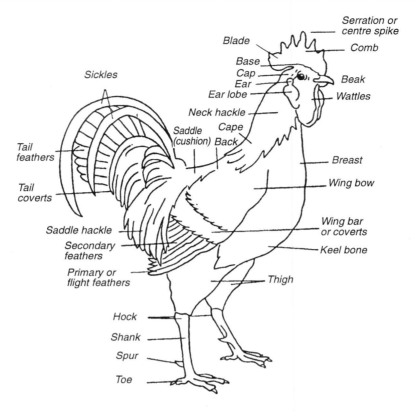

Figure 2.1   Points of a chicken. Alternative names apply to the female. *(Based on M. Whalley-Taylor)*

way of China and Central Asia to Siberia, Russia and the Near East, and secondly, via Persia and Greece to the southern areas of Europe.[4] The Persians are thought to have introduced them from India following its conquest by Cyrus in 537 BC. Two hundred years later, Alexander the Great conquered Persia and the chicken found its way to Greece as the 'Persian fowl'.[5] When Julius Caesar came to Britain in 55 BC, he found that the custom of cock fighting was well established, probably introduced by the trading Phoenicians. It is thought that these fowl were similar to Old English Game.

'They think it wrong to eat hares, chickens or geese, keeping these creatures only for pleasure and amusement.'[6]

The Roman conquest is thought to have introduced heavier birds into Britain. In AD 47 Columella described a five-toed and broad-breasted bird, in Rome, which was similar to our present Dorking, a breed that differs from most others in also having five toes:

'Let the breeding hens be of a choice, of robust body, square-framed, large and broad breasted, large heads, with small erect combs and white ears – those hens are reckoned the purest breed which are five-clawed.'[7]

The practice of caponising and rearing for the table was probably also introduced by the Romans who had inherited it from the Greeks. It was a common practice in Rome, judging by Varro's treatise *On Farming*:

'The term hen is applied to female

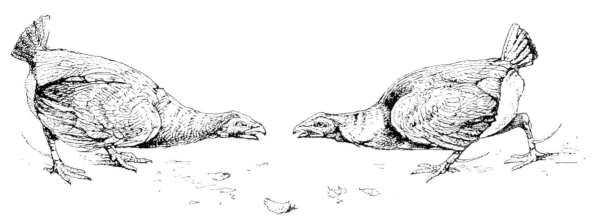

Figure 2.2   For centuries, the main interest in poultry was in cock fighting. *(From an engraving by Scott after a painting by Berango, British Field Sports, 1818)*

barn-door fowls; the males are called cocks, the half-males – those which have been castrated – capons.'[8]

Welsh records indicate that the practice of caponising was probably widespread in pre-Saxon Britain. An early Welsh record of an older oral tradition tells of an immense feast staged by Caswallon, son of Beli, in Llundain (London) where 50,000 geese and capons appeared on the menu.[9] English references to capons are to be found in Shakespeare and in a seventeenth century treatise *Cheape and Good Husbandry:*

'If you will have fat crammed chickens, you shall coop them up when the Dam forsaketh them and the best crams for them is wheatmeal and milk.'[10]

Historically then, chickens were regarded as table fowl or fighting cocks and there was comparatively little emphasis on egg production. Selective breeding for fighting capabilities certainly took place, as indicated by Gervase Markham's advice in 1614:

'In your Election chuse him which is of a strong shape, good colour, true valour, and of a most sharp and ready heel.'[11]

It was usually the task of women to feed the chickens, although men ran the cock fighting. After cock fighting was declared illegal in the UK in 1849, chickens continued to be the province of women. They provided 'hen money' at a time when, according to Edward Brown, *'there was a contempt for poultry as a branch of livestock – one that was only worthy of the attention and endeavour of women'.*[12] The publication, in 1886, of *The Hen Wife*, written by the Honourable Mrs Arbuthnott, appears to continue the trend: *'It is impossible to imagine any occupation more suited to a lady living in the country, than that of poultry rearing'.*[13] There were echoes of this still to be found as recently as the early 1950s, in the practice of the farmer's wife having the 'egg money', while the farmer carried out the serious business of farming!

Another factor which undoubtedly delayed the development of commercial poultry farming in Britain was the attitude of the landed gentry, who preferred large farms because this facilitated their hunting. One of them actually declared that *'poultry can only be kept on sufferance'.*[14] Attitudes such as this arguably contributed to the fact that America rather than Britain was eventually to play the leading role in commercial poultry development.

From the nineteenth century onwards, three major factors were to have radical influences. The first was the introduction of the large Asiatic breeds that carried the genetic factor for brown egg shells. These included the Cochin, Brahma and Langshan. The Cochin caused a sensation when it first appeared in Britain in 1843. The sheer size of the breed and the fact that a trio was presented to Queen Victoria ensured its popularity. (Royalty has maintained its connection with poultry breeds and with the Poultry Club of Great Britain to this day.)

Interest in the Cochin and the other large Asiatic breeds fostered a huge and general interest in poultry, as well as in the staging of poultry shows. Earlier shows had been village affairs with the old cock-fighting 'cockers', but in 1845 the first real Poultry Show took place in London. In 1863 the first Poultry Club was formed, while 1874 saw the launch of *Poultry World*, a publication which continues today as the monthly magazine for the poultry industry. Until the period of intensive poultry farming began in the late 1950s, some of the national daily papers had poultry columns.

Until the introduction of the Asiatic breeds, most eggs were white or slightly tinted. Crossing existing breeds with the new introductions resulted in the brown egg shell factor becoming much more widespread. An indication of the interest in breeding was the invention in Britain of the first effective small incubator, patented in 1881 by Hearson. The first Laying Trials started in 1897.

The second factor in the overall development of breeds was the introduction of the light Mediterranean breeds such as the Leghorn, Ancona and Minorca. These, particularly the Leghorn, were prolific layers of white eggs and their influence on other breeds was to dramatically improve their productivity. For the first time, numbers of eggs became important and, with appropriate breeding, this factor could also be introduced to brown egg layers.

Figure 2.3   Old English Game, the type probably brought into Britain by the Phoenicians, and which Julius Caesar found when he invaded. *(Illustrated by Harrison Weir in 1874)*

In 1880 William Cook, of Orpington in Kent, created the Black Orpington, arguably the first of the dual-purpose breeds. It laid a good number of brown eggs, while also being suitable for the table. He bred the Black Orpington by crossing a black Minorca male with a black Plymouth Rock hen. The pullets from this cross were then crossed with a Langshan male. The Black Orpington was also exported to Australia to become the Australorp, and established a good record of production there. In Britain, however, what had been an excellent utility breed was soon turned into a show bird.

The third factor was the American influence, which took the best utility breeds and not only improved them scientifically, but also developed new, dual-purpose breeds. The fact that the Americans invented the trap-nest, a device for recording the number of eggs laid by specific hens, perhaps

A range of breeds in the last century (above) and of traditional breeds today (below).

indicates their commitment to scientific development. It was not until 1902 that it was introduced and first used in Britain, although a Utility Poultry Society had been formed in 1896.

The Rhode Island Red, which arrived in Britain in 1904, is a supreme example of a highly developed, dual-purpose breed. It was to have an immense influence on future breeding, as well as being a popular choice with show breeders. Yet my grandfather, as a young man, remembered that when it first appeared it was regarded as an American mongrel!

The Rhode Island Red was widely crossed with other breeds such as Light Sussex in order to produce commercial crosses. With a Light Sussex hen it produced a first cross that was sex-linked. In other words, the chicks could be identified as male or female by differences in down colour. The silvery yellow males could then be raised as table fowl, while the brownish females were kept as replacement egg layers. It was also the bird that was to give rise to all the commercial brown egg layers that we have today, just as the Leghorn provided the major genetic source for white egg layers. The Americans also considerably improved the productive capabilities of the Leghorn and other Mediterranean breeds.

New, more productive utility breeds were certainly needed in Britain at that time, although the interest in show birds was still high. With few exceptions, the stock was quite unable to meet consumer demand. As

Figure 2.4   Old type Dorkings illustrated by Harrison Weir in 1853. This breed was probably introduced by the Romans, who also brought the practice of caponisation.

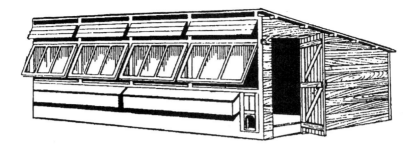

Figure 2.5   Commercial layer's house manufactured by Boulton & Paul of Norwich in 1924. Note the large windows to maximise natural daylight and ventilation, although the pop-hole is small by today's welfare standards.

late as 1914, nearly 18 million eggs had to be imported into Britain, some from Russia. World War I gave an impetus to home production, and commercial strains and crosses of Rhode Island Red, Light Sussex, White Leghorn, Wyandotte and Plymouth Rock were increasingly kept. In 1923, a Mr P. A. Francis of the Ministry of Agriculture was *'despatched to study poultry conditions in the United States and Canada'*, according to *Feathered World Yearbook*, 1924.[15]

Some of the new poultry farmers were ex-servicemen who had managed to survive the death fields of Europe. Firms such as Boulton & Paul of Norwich produced houses that met the needs of the new enterprises. In an advertisement of 1924 it offered *'a commercial egg house specially designed for making poultry profitable – correctly lighted and ventilated, draught and damp-proof – perches and trap nests correctly placed – and all parts easily accessible for cleaning.'*

Wyandottes and Plymouth Rocks were the breeds kept by my parents on their Welsh smallholding in the 1920s–40s. Their houses were not as sophisticated as Boulton & Paul's, as they were knocked up by my great-uncle who was the village carpenter. My parents created great interest in our small village by being the first people to order, and have delivered, a consignment of day-old

chicks, an unheard-of practice in that remote area. Half the village turned out to see the arrival of these English importees from Mytholmroyd, and there was a steady stream of visitors to examine the paraffin brooder and its exotic inhabitants. Despite dire predictions to the contrary, they proved to be excellent layers.

Some interesting work on the development of autosexing breeds was carried out in Cambridge from 1929 to 1941. Normally, sex-linkage, where chicks can be identified as male or female at hatching, requires two distinct breeds, such as the Rhode Island Red × Light Sussex cross, referred to earlier. An autosexing breed, by contrast, is one breed, such as the Cambar, which produces identifiable sexes at hatching. Crossing a Gold Campine with a Barred Plymouth Rock, followed by judicious selection, produced the Cambar in Cambridge, while apparently revealing the principle of barring as an autosexing device to Professor Punnett and Mr Pease. Whether this was an accidental discovery is not clear. No doubt an examination of the original papers, some of which are apparently in French, will reveal the details.

Autosexing breeds clearly had a commercial application, but the emphasis on hybridisation after World War II effectively put a stop to this line of research. It has continued to a limited degree in the non-commercial sector, with various autosexing breeds now being recognised by the Poultry Club of Great Britain.

In addition to halting the work on autosexing breeds, World War II brought conscription, shortages, rationing and dried eggs. It also undoubtedly boosted the keeping of backyard poultry by families who wanted to have fresh eggs.

After the war, many ex-servicemen used their gratuities to start egg farms, just as an earlier generation had done. The demand for fresh eggs was there after years of rationing, and the Ministry of Agriculture now instigated a new poultry department to provide information and advice. Books on how to start a poultry farm were printed by various publishers, but many were out of date almost as they became available. The poultry industry moved forward at a speed previously unknown. It quickly diverged into two branches – egg and broiler production. Hybrid birds were developed either as high-yielding egg strains for the battery cages, or quick-growing broiler strains for environmentally controlled rearing houses. *Poultry World*, reflecting these changes, became the magazine of the intensive poultry industry, although still giving some coverage to the show sector.

Many smallholdings quickly diversified into battery egg units or market gardens, or were sold off. The traditional family smallholding became an endangered species. Many people, my parents included, moved to the cities. No longer with a readership, the weekly *Smallholder* magazine changed itself into a gardening publication called *Popular Gardening*. This was eventually absorbed by *Amateur Gardening* magazine.

Throughout the 1950s and 60s intensive farming boomed, and Britain for the first time was able to bring cheap, fresh eggs and chicken meat to the tables of the entire population, albeit from batteries and environmentally controlled rearing houses.

The older breeds of poultry were commercially redundant. Many of them came close to extinction but were preserved by the dedication of small breeders, many of them pursuing poultry keeping as a spare-time hobby.

In the late 1960s and 70s a reaction against what was perceived as gross intensification and consumerism began. It created a self-sufficiency boom that had thousands moving from the towns to the countryside. My husband, David, and I were part of that migration. As there was no magazine for smallholders, or any information relating to free-range poultry and small-scale farming, we started our own publication. The bi-monthly *Practical Self Sufficiency* was launched in 1975, changed its name to *Home Farm* in 1983, and continues today as the monthly magazine *Country Smallholding*. In 1981 *Fancy Fowl* magazine was started by Shirley Murdoch for those with show poultry. After a quarter of a century gap during the intensive farming boom, small-scale poultry keepers, utility and fancy, were being catered for again.

In the self-sufficiency period, free-range poultry keeping was resurrected and with it a demand for chickens. Many, myself included, bought traditional breeds of large fowl, such as the Rhode Island Red, only to find that their productive capabilities had, by now, diminished at the expense of show characteristics. For the first time, hybrids bred for the batteries were kept in free-range conditions. It became almost fashionable to 'rescue' battery hens, buying them at the end of their commercial lives (one laying season) and teaching them to walk again.

Consumer reaction against battery conditions and a concern for more wholesome foods had also created a considerable demand for free-range eggs. The first large-scale free-range units were set up, but the battery mentality prevailed in some. Abuses such as not allowing the flock out for most of the day resulted in complaints that this was not real free-range.

Legislation was introduced in 1981 to ensure that such abuses did not continue, and the term 'free-range' became a trade description.

As with all legislation, it was a two-edged sword. Some small poultry keepers found that they had to stop selling their surplus eggs as free-range ones, even though the conditions they provided were exemplary. They had to use other descriptions such as 'fresh eggs', at a time when battery producers could use a wide range of euphemisms such as 'fresh from the farm' with impunity. Even now, eggs from caged birds do not have to be described as battery eggs, but those from alternative systems have to use one of four descriptions: free-range, semi-intensive, deep litter, perchery (barn), depending on the management system.

In recent years, the RSPCA has brought in its own highly successful standards for producers to follow, enabling those who are registered with it to use the Freedom Food designation as a supplementary description. Well before this, the Soil Association and other organic bodies had drawn up humane standards. In 1987, these were brought under the auspices of UKROFS.

In 2000 the EU standards were implemented.

Throughout the 1980s, interest in traditional breeds continued, small poultry keeping was practised on a widespread scale and free-range produce continued to make headway into the multiple stores. Then, just before the first edition of this book was published, in 1990, Edwina Currie, in the Health Ministry, made her famous comment that salmonella was present in all sectors of the poultry industry. It was quite true, but she made the political error of putting the interests of the consumer before those of producers. It created an enormous and short-lived reaction against eggs, but the serious consequence was the ill-thought legislation rushed in by the Government.

Owners of all flocks of 24 birds or more were suddenly required to test their birds regularly, and so were tiny domestic flocks, if any surplus eggs were sold. The effect on large enterprises was minimal, but on small flocks it was disastrous. Many people stopped keeping chickens and some irreplaceable and old-established utility flocks were dispersed and lost. Coinciding, as it did, with an economic recession, it also brought about the demise of many suppliers of poultry housing and equipment. Pressure was brought to bear on the Government from many quarters in Britain and Europe, as some traditional breeds teetered close to extinction.

The legislation was eventually amended and brought into line with the European Union so that, now, only breeding flocks of 250 birds or more are required to undergo salmonella testing. Laying flocks and small breeding flocks are exempt. These changes have encouraged many people back into poultry keeping, and with them an increased chance of survival for the traditional breeds. It is to be hoped that our legislators will refrain from further ill-thought action that will damage the small poultry sector.

In 1992, the Rare Breeds Survival Trust launched a Poultry Support Project. This is a programme to increase the numbers and productivity of some traditional utility breeds. They include Indian (Cornish) Game, Derbyshire Redcap, Scots Dumpy, Scots Grey, Croad Langshan, Old English Pheasant Fowl, Dorking and Sussex. The programme includes setting up accredited units where production records are kept, and breeding from the best birds in each case.

At the time of writing, small poultry keeping is again booming and there is a renewed interest in pure breeds. The Poultry Club of Great Britain and the Rare Poultry Society continue to work for the interests of this sector. EU regulations regarding transport of livestock have made the delivery of small consignments of birds difficult, but private delivery services and individual collections at farms and shows are coping with the situation.

On the commercial side of free-range production, the outlook is bright, although some elements of the large-scale sector are still using barbaric practices such as electrified cables inside the house to prevent floor-laid eggs. Most birds for the large sector are also routinely beak-trimmed, although those kept by organic producers are left intact. Supermarket chains are selling organic and free-range eggs, as well as free-range table poultry. Some have even instigated their own standards which ban beak trimming.

Perhaps the most surprising development is the appearance of new hybrids, such as the Speckledy and Columbian Blacktail, which are described as 'genetically naive' because they are similar to birds kept in the pre-hybrid period. These are to provide speciality eggs for the free-range sector. The breeders have had to go back to the genetic pool of the pure breeds to develop them. It can be said, with some justification, that history is repeating itself, for all the commercial hybrids owe their existence to the traditional utility breeds.

Characteristics

Our present-day *Gallus domesticus* is bigger, more productive and has a wider range of colour and external attributes than her early ancestors, but she is still essentially the same in the way she behaves. The chicken is a pecking, scratching, perching creature, subject to the pecking order of its peer group. It likes to flap its wings and take dust baths. Having a clear knowledge of what these characteristics are, and how they manifest, enables us to provide adequately for its well-being. Although the chicken is a highly developed entity, it does not need to be treated as a machine, as it is in the battery sector.

Non-oiled feathers

Chickens originated in a warm jungle environment, with plentiful cover against the elements. Unlike ducks, which have a well-developed preening gland, chickens have only a rudimentary one. As a result, their feathers are not well oiled and therefore have no proofing against the weather. Some breeds are more close-feathered than others, but it is not true, as some have claimed, that pure breeds can stand up to the rain. Any chicken in the rain will become wet. A sturdy house with a protected scratching area is essential. So too, is the provision of outside shade and security.

Fear of predators

A chicken is prey, not predator, hence its monocular vision, with eyes on either side of its head to have better all-round sight. It is obvious that it needs to be protected against enemies such as foxes, dogs, cats and rats. What is not always understood is that it also needs the security of cover; otherwise it is reluctant to range very far. Again, it is harking back to its jungle origins. Trees, hedges and sides of buildings are ideal, while shelters which it can duck under are also advisable. To a chicken, even a passing plane is an overhead predator.

Perching

Like those of most birds, the chicken's feet and legs are adapted for perching above ground. Any house should be equipped with perches, and new birds should be purchased from breeders who raise ground-reared stock which has had access to perches. They will then adapt to a new environment without any problems.

The feet and legs are well adapted for their perching or roosting function. The toes grip the perch and the muscular system acts rather like an automatic lock so that the bird does not fall off while it is asleep. The perching instinct is a survival technique, giving protection against predators who, in the original environment, would roam the jungle floors while the fowls perched on tree branches.

Pecking

The chicken pecks its food with a beak that is well designed for the purpose. It is capable of picking up a single grain of wheat or of snipping off the growing tips of young grasses. It needs, therefore, to have food that is appropriate in size and texture to its needs. Specially formulated feed pellets or mash and grains are suitable, but sunflower seeds or maize grains need to be crushed or kibbled (chopped) before feeding. Similarly, fresh, young pasture will be appreciated, while old, rank growth will not.

The commercial practice of beak trimming (removing the tip of the upper mandible) is incompatible with the bird's instinct. How can a chicken be described as a free-ranger if it is unable to glean effectively? I conducted a small experiment where beak-trimmed hybrids were put on an area of concrete. Some layer's pellets were then strewn thinly on the ground. Most of the birds were unable to pick them up. I then repeated the same experiment with the birds on grass. This time, they were able to pick up the pellets, presumably because the softer ground provided sufficient 'give' for the beak to get round them.

The justification for beak trimming is that it is necessary to stop feather/vent pecking and cannibalism that sometimes occur in overcrowded conditions. Where chickens are housed and managed in smaller flock densities, there should be no need to resort to such practices. If a system is to be described as free-range, it should be so.

Breeding companies producing birds for the free-range sector have produced more docile strains in recent years. Hopefully they will continue to select birds for

non-aggression so that beak trimming will soon become a thing of the past.

The pecking order

The existence of a pecking order does not mean that unless beak trimming is allowed, the chickens will peck each other to death. Domestic chickens are social creatures, with a well-established order of hierarchy. It is a hierarchy in which some birds are more dominant than others. The cock is usually 'cock of the walk', but in the absence of a male, an older hen or even the poultry keeper who may be regarded as the dominant force will take on this position. Anyone who has gone into a poultry house or area where there is a flock of female chickens, particularly modern hybrids, will have seen the tendency that many have of squatting in a submissive way, or even allowing themselves to be picked up without fuss. This can be a positive advantage where removal of birds is necessary.

The negative side of the pecking order is that some birds may bully others. It is important to watch out for this, particularly where docile birds are prevented from feeding adequately at the troughs; otherwise egg production, for example, will decline. It has been demonstrated that competition for food increases the incidence of aggressive pecking.[16] There should be enough feeders to ensure that crowding does not take place, and pop-holes should be sufficiently wide and numerous to cater for stress-free exit and access. When I come across the occasional bully, my practice is to separate the bird for a few days of solitary confinement in a small house and run. She can still be seen by the rest of the flock so that when reintroduced, she is not treated as an outsider. This normally has the effect of disorientating the culprit to the extent that the original pattern of behaviour is forgotten.

Scratching

The feet, with their long toes and sharp

claws, are excellent for scratching, while the powerful muscles of the legs wield considerable force. On free-range they will use these abilities to the full, and it is a good idea to give them a scratch feed of grain in the afternoon. They can then exercise two instincts at once – scratching and pecking. This is also an effective way of getting them to range further away from the house so that the area immediately around the building does not become over-used and damaged.

On a small scale, they are no respecters of flower and vegetable beds. It is necessary to confine them to a run, orchard or field where they cannot do any damage. By the same token, their scratching activities can be put to good effect in the vegetable garden after all the produce has been harvested. Putting a moveable house and run on an area that requires weeding and clearing is an effective way of preparing beds for spring planting. It also ensures that slugs and other pests are controlled.

Dust bathing

This refers to the pattern of behaviour where a chicken scratches an area of fine, dry soil and then takes a 'bath' in it, allowing the soil to trickle through the feathers. It is an instinctive action to help get rid of external parasites in the feathers and on the skin. Most free-ranging birds will normally find a dry, sunny spot where they can make their own dustbaths, but if they are in a run where this is not possible, a large shallow box with fine sand should be made available to them. Ideally this should be under cover.

If an outbreak of lice or mites does affect a flock, it is not sensible to rely completely on a dust bath. The individual birds, the houses, perches and the dust baths themselves should be treated with an appropriate product.

Egg laying and broodiness

Like all birds, chickens lay eggs. Where they differ from most birds is that they lay far more. Originally, they laid two or three clutches of eggs, depending on the season. Each clutch would contain about ten to twelve eggs and, once these were laid, the female would sit on them, in order to incubate them, for twenty-one days, until they were hatched. This is the pattern followed by all birds in the wild, although the period of incubation differs, according to the type and size of the bird. Chickens have been selectively bred in order to lay far more eggs than their ancestors laid, but they have still retained all the instincts of their wild forbears and should be catered for accordingly. In other words, they need a quiet, darkened area where they can lay eggs in peace. Wood shavings or other nesting material is required to line the nest boxes.

There are occasions, of course, when broodiness is not required. Modern egg laying hybrids have, to some extent, had broodiness bred out of them, but it can still happen. Warm weather can trigger it, as well as seeing a cluster of uncollected eggs. It is also more prevalent in hens in their second year.

It is necessary to remove the broody bird and put her in a cool, well-ventilated coop. The drop in temperature and absence of eggs will soon break the broodiness pattern, allowing her to rejoin the laying flock.

Moulting

This is a process by which old feathers are shed, to be replaced by new ones. It is an annual event, starting when the bird is a year old, and usually takes place in the summer or autumn, depending on time of hatching. It is quite natural, but some who have never seen it before may assume that something is wrong. In domesticated chickens autumn- and winter-hatched birds will moult between July and August. Those hatched after March will normally continue through to October or November before moulting starts. It usually lasts a few weeks and egg production declines, and may even cease, while it is going on.

Moulting is also linked to feed patterns and stress, and it is important not to force birds to be more productive than their metabolisms can cope with. This is discussed further in the Poultry Management chapter.

Clucking and crowing

Cocks crow loudly, particularly at daybreak. It is a territorial call and many country people regard it as an integral part of country life. Some, however, have a different attitude. The keeping of a cock should only be considered by those unlikely to have near neighbours, and only if serious breeding is to be considered. Cock crowing is a recognised 'sound nuisance' in law and there have been numerous incidences where owners have been required to get rid of one. In Germany there is a tradition of cock-crowing competitions, using Bergse Kraaier birds, to see which has the longest or loudest crow.

There is no truth in the old belief that hens lay better when there is a cock. In fact, the opposite is the truth. Hens can be damaged by the spurs of the male and infections such as vent gleet (poultry venereal disease) can also be transmitted. Any eggs which are to be sold should come from hens which are not kept with a male.

Hens also have a range of sounds, although they are not as vocal as the male. A single hen who finds an object of interest while scratching talks to herself with quiet little clucks. A mother hen with chicks gives a similar sound, but louder. Hens produce a slightly complaining, cawing note just before laying an egg, and once the egg is laid, there is a triumphant cackling. A single chirring note is a warning call in the event of danger, or if a stranger to the flock is sighted. The cock also produces a series of staccato chirring sounds when he is about to attack.

There are many other vocal variations and it would make an interesting study to try to decipher the various sounds. The small poultry keeper is probably in the best position to do this because there is more time to develop a relationship with individual birds than would be the case with a large commercial flock.

References

1. Dr Akishinonomiya *et al*, *Proceedings of the National Academy of Sciences*, 1994.
2. Author's visit to the tomb of Tutankhamun, Valley of the Kings, Egypt, 1983.
3. *The Bible*, Matthew 26. 34.
4. Victor Hehn, *Wanderings of Plants and Animals*, 1895.
5. Edward Brown, *Races of Domestic Poultry*, 1906.
6. Julius Caesar, *Gallic Wars* 5, 12–14, Anne and Peter Wiseman.
7. Columella, AD 60–65, *On Agriculture*, Lewis and Rheinhold.
8. Varro, 37 BC, *On Farming*, W.D. Hooper.
9. Rachel Bromwich, *Trioedd Ynys Prydein (The Welsh Triads)*, University of Wales, 1978.
10, 11. Gervase Markham, *Cheape and Good Husbandry*, 1614.
12, 14. Edward Brown, *British Poultry Husbandry*, Chapman & Hall, 1930.
13. The Hon. Mrs Arbuthnott, *The Hen Wife*, 1886.
15. *Feathered World Yearbook*, 1924.
16. M.G. King, 1965, 'Disruptions in the Pecking Order of Cockerels Concomitant with Degrees of Accessibility to Feed', *Animal Behaviour*, 13, 504–506.

CHAPTER 3

Breeds

In order to make a success of any branch of poultry-keeping, it is important to keep the right breeds.

Herbert Howes, 1939

The world 'breed' is often used quite loosely to denote a certain group of birds with similar characteristics. There is more to it than that, depending on the genetic makeup of the bird. As far as type is concerned, some are better than others for particular applications. Before we take a detailed look at the question of breeds, here is an overview of what *types* of birds are the appropriate choices for the various branches of poultry keeping.

Table 3.1 Appropriate breeds for specific use

BEST EGG PRODUCTION
Type:**Modern hybrid**

ISA Brown	Babcock 380	
Lohmann Brown	Black Rock	Hebden Black
Hisex Ranger	Bovans Nera	Bovans Gold
Hisex Brown	Speckledy	
Shaver 579	Columbian Blacktail (Calder Ranger)	
Hy-Line Brown	Bovans White (White Star)	

BEST TABLE BIRDS
Type: **Modern hybrid**

White-feathered		*Colour-feathered*
Cobb	Ross	Hubbard ISA Redbro
Hybro		(Poulet Bronze)
Sherwood White		Sasso 551 (Poulet Gaulois)

BEST FOR SHOW
Any recognised pure breed of large fowl or bantam (see Table 3.6)

BEST FOR GENERAL INTEREST
Any chicken you like

Development of breeds

Figure 3.1 gives a rough indication of the overall development of chickens, but as it is an oversimplification, it should be regarded only as a general guide. It shows that there was an early division into large fowl and bantams. When and how this happened, and whether they had different ancestors, is unknown. What is known is that man has since developed miniature strains of large fowl that are now also called bantams.

What is also apparent from the plan is that there is a considerable variation in the types of fowl, including the availability of heavy, light and dual-purpose breeds.

Heavy, light and dual-purpose

The heavy breeds are large and therefore more appropriate to the table. They are poor fliers, therefore easy to confine, and have a greater tendency towards broodiness (the desire to sit on and incubate eggs). Because of this, they are also referred to as *sitting* breeds. Examples are Dorking, Maran and Indian Game.

Light breeds are smaller and have a greater tendency to flight. They also produce more eggs than the heavy breeds and have less of a tendency to become broody. Examples are Leghorn, Ancona and Welsummer.

Dual-purpose breeds are those that have

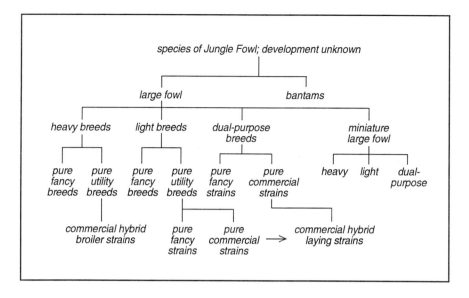

Figure 3.1   A rough guide to the development of chickens.

characteristics of both heavy and light breeds, in that they are of reasonable table weight and also provide a good number of eggs. Because of this, they were often the choice of poultry farmers until the period following World War II. Examples are Rhode Island Red, Light Sussex and Wyandotte.

Utility and fancy

Breeds have also diversified into *utility* and *fancy* strains. Utility are those that have (or had) good production records for eggs or table, or both. Fancy are those that have been selected for show appearance rather than production levels. Some breeds, of course, have never been renowned for their utility characteristics. An example is the Sebright, which was developed purely as a show bird.

Pure breeds, crosses and hybrids

A pure breed may be large or bantam-sized, heavy, light or dual-purpose. The salient point is that it is genetically standardised as a breed and will always breed true. In other words, the birds will produce offspring exactly like themselves when crossed with each other. Examples are the Maran and Dorking.

A first cross is a bird that has been produced by crossing two different pure breeds, such as Rhode Island Red male × Light Sussex hen.

Hybrids are the product of several different breeds, the progeny of which demonstrate particularly good characteristics such as number of eggs laid, size and quality of eggs, etc, and are selected for further breeding. A modern hybrid is the result of these various strains being brought together, utilising genetic material from these different sources. They are highly bred for efficient production. Some free-range hybrids are now able to produce over 300 eggs at 74 weeks.

Most of the large-scale selective breeding, as we have seen, took place in the

The author's Cuckoo Maran cock with a sex-linked Rhode Island Red × Barred Plymouth Rock hen.

however, the chick markings are quite different. The females acquire a single dose of the barring and have a distinct light patch, while the males, with a double dose, have the light patch extending down the back. As a result, they are altogether lighter and easily distinguished.

The original Cambar was made from a gold Campine and a Barred Rock. The salient factor about an autosexing breed (as distinct from a sex-linked cross) is that only one breed is involved. In other words, two Cambars will produce identifiable male or female chicks. They, in turn, also have the autosexing characteristic. A sex-linked cross involves two different breeds, such as Rhode Island Red and Light Sussex. Several autosexing breeds have been developed, and are indicated in Table 3.2.

commercial poultry world, with the emphasis on either high-yielding egg strains or quick-growing broiler strains. These are the ones chosen by anyone with a commercial interest, as their production records are better than those of the more traditional breeds. Unlike pure breeds, they will not breed true. This does not mean that they should not be crossed, merely that the characteristics of the resulting chicks cannot be accurately predicted. Some small poultry keepers cross some of their best hybrid layers with a pure bred male such as a Rhode Island Red. The problem is in finding one from a good productive line.

Autosexing breeds

Mention was made in the previous chapter of the autosexing breeds first bred in Cambridge. Breeds such as Barred Plymouth Rock and Maran carry a genetic factor for barring, indicated by a light patch on the head of day-old chicks. When crossed with black-feathered breeds, all the chicks have a light head patch and the sexes cannot be distinguished. When the barring is transferred to brown-feathered breeds,

Table 3.2 Autosexing breeds

Cambar	Barred Campine
Legbar	Barred Brown Leghorn
Dorbar	Barred Dorking
Brussbar	Barred Brown Sussex
Wybar	Barred Lace Wyandotte
Brockbar	Barred Buff Rock
Welbar	Barred Welsummer
Rhodebar	Barred Rhode Island Red

It is important to remember that some breeds which appear to be pure may not necessarily be so. They may carry genetic factors which interfere with the transference of the barring so that chicks do not necessarily carry the distinguishing factors. Merely crossing one breed with another will not produce an autosexing breed. Several generations of painstaking selection are needed. All breeding requires a great deal of patience, careful recording and a determination not to breed from failed experiments.

Bantams

Originally, bantams were probably naturally occurring dwarf forms of chickens resulting from random and spontaneous genetic mutations. As the resultant birds bred, the dwarf or bantam characteristics were passed on to the offspring. No one is certain where they originated from, although Java has traditionally been regarded as the most likely possibility. By the beginning of the nineteenth century, there were several different forms in Britain, including the Java bantam, the English dwarf and the French dwarf. Other breeds are Pekins, Nankins, Rosecombs, Sebrights and Booted bantams.

In addition to true bantams, which have no large counterparts, scaled-down versions of large fowl have been selectively bred by poultry breeders. These are now usually referred to as bantams although, to be accurate, they are miniature fowl. Examples are available from many of the large breeds, and as miniature forms are developed they are standardised and recognised by the poultry organizations. Examples are Rhode Island Red, Light Sussex, Ancona, Hamburgh, Barnevelder and Maran.

Silkies, which are sometimes thought of as bantams, are in fact large fowl that are smaller than the average large fowl. Of Asiatic origin, with silky feathers, they have feathers on the legs.

Many people keep bantams for interest, rather than for their utility value. They have not been bred for egg production and will not produce the numbers of eggs that commercial hybrids will. Nevertheless, where space is limited, such as in a small garden, bantams may be more suitable than large fowl. Their other advantages are popularity with children and reduced feed requirement. It has been estimated that the average intake is less than one-third that of a large fowl, but there is naturally a variation depending upon size, age and level of activity. Bantam eggs are approximately half the size of normal large eggs, but this is a generalisation for there is considerable variation depending on the breed, the age and the level of feeding.

Perhaps the greatest attribute of bantams generally is that they make excellent foster mothers. Many of the bantam breeds, particularly the Pekin, have a strong tendency towards broodiness several times a year and often early in the season. For this reason, they are particularly popular with pheasant breeders, and those who wish to raise replacement stock by natural incubation. Not all bantams are equally good when it comes to mothering qualities, of course. Some, such as the Sebright and Old English Game, can prove difficult, showing no inclination to become broody. It should also be remembered that individual birds vary in ability to become broody, regardless of breed and type.

Breeds for egg production

If the aim is to produce the maximum number of eggs possible under free-range management, the choice of bird will be one of the modern hybrid strains of large fowl bred for brown or white egg production. These are the ones indicated in the table at the beginning of the chapter. The brown egg layers are based on the Rhode Island Red, while the white egg layers are related to the Leghorn.

British consumers prefer brown eggs, perhaps seeing them as more wholesome and associated with the farm. There is no truth in this belief! The factor which determines shell colour is a genetic one. If a hen which lays white eggs, such as the traditional Leghorn, is kept on free-range, she will continue to produce white eggs, just as a brown layer will produce brown eggs in a battery cage.

While it is certainly true that in egg laying trials in the past the Rhode Island Red was recorded as having produced over 300 eggs in extensive conditions, it is no longer able to do so. As referred to earlier, most of the pure breeds that are available now have been bred for show, not for their productive capabilities.

Some suppliers sell hybrids based on traditional first crosses. The popular Black Rock, for example, is derived from the Rhode Island Red and Barred Plymouth Rock cross. It has been bred for hardiness, in addition to egg laying capacity, and is a good choice for free-range conditions, particularly in more exposed areas of the country. The black-feathered Bovans Nera, which looks almost identical, is based on the same cross. In America, it is called the Black Sex Link. It is important to bear in mind that merely crossing any Rhode Island Red male with any Barred Plymouth Rock female will not produce a good commercial cross. It must be developed from selected strains that have shown good production levels.

Modern hybrid layers have been largely bred for the battery industry, with the emphasis on breeding a light bird in an effort to improve feed conversion, but experience over the last decade has shown them to be quite capable of adapting to free-range, although no chicken can cope with cold, wet and exposed weather conditions. Recent findings do, however, indicate that pullets destined for free-range conditions need to be heavier than those going into cages; otherwise they may not be able to eat enough for body metabolism, energy requirements and good-sized egg production. Natural floor rearing conditions without giving too much extra light will ensure proper development without giving rise to precocious birds.

In the last few years, several poultry breeders have developed slightly heavier hybrids for the free-range sector. Examples are Lohmann Brown, Hisex Ranger and Bovans Nera which show good production rates in extensive conditions.

Table 3.3   Weight differences between battery and free-range birds

	Battery Bovans Brown	Free-range Bovans Nera
17 weeks	1360–1400g	1400–1460 g
18 weeks	1470–1530g	1490–1560 g
20 weeks	1640–1740g	1670–1770 g

Source: Hendrix Poultry Breeders, Bovans Fact Sheets, May 1996.

In the first edition of this book, I commented that where hybrids had not yet been able to compete with some of the old breeds, such as Maran and Welsummer, was in the production of very dark brown, speckled eggs. Since then, the Speckledy, a hybrid based on the Maran, has been introduced for this sector of the market. Its eggs are certainly attractive, and have found their way into the supermarkets, but the best mahogany, speckled eggs are still to be found in good examples of the pure-bred Maran. Since the introduction of the Speckledy, some Maran breeders have been crossing it with their stock in order to increase productivity, while retaining quality colour and markings. It should be remembered, however, that a breed such as the Maran or the Welsummer will produce beautifully dark eggs only if the bird is a good example of the breed. Inferior

A Lohmann Brown hybrid.

The Speckledy, a new hybrid layer, based on the Maran, to produce brown speckled eggs in commercial numbers.

obvious advantages to having the supplier on the door-step, and if any problems arise, they can be readily sorted out through direct contact. Some packers and franchisers advise on which hybrids are kept by their producers. What is important for free-range layers to establish is the production record and the egg weight profile. The former will be evidence of a bird's parentage and productivity, indicating the number of eggs it is likely to produce. These factors are crucial to a free-range producer. The ideal free-range egg layer has the following characteristics :

- Adequate body weight at start of lay
- A good hen-housed average (HHA) for the number of eggs laid
- A number of different egg sizes, none too large
- A moderate feed intake in relation to its size

A hen-housed average is the total number of eggs produced by a flock, divided by the

specimens will lay eggs that are not much different from any other brown egg. Top-class examples really are almost chocolate brown, with ample speckles, and can command a premium price.

Another recently introduced hybrid for the free-range sector is the Columbian Blacktail. This also lays speckled eggs, not as marked as those of the Speckledy, and production is said to be slightly better. The Columbian Blacktail is a breed name that is exclusive to Waitrose free-range producers. Elsewhere the breed is called Calder Ranger.

The choice of bird may be largely decided by which supplier is most conveniently placed for a particular area. There are

Some of the author's ISA Browns.

number of birds (including mortalities). This may be worked out at any number of weeks (age of birds) so that some breeders may be quoting an HHA for 70 weeks, while others may be basing it on 74 weeks. Watch out for this!

There are other considerations to be taken into consideration, as the following table indicates. This gives the expected performance data for the Hisex Ranger, bearing in mind that management and environmental conditions may vary.

Table 3.4   The Hisex Ranger perfomance data

Rearing period	Livability	97.0%
	Body weight at 17 weeks	1500 g
	Feed consumption per bird	6.34 kg
Laying period	Egg production, hen housed	292
(to 72 weeks)	Age at 50% production	150 days
	Average egg weight	63.2 g
	Egg mass	18.5 kg
	Feed consumption	120–130 g/bird/day
	Feed per egg	144g
	Feed conversion ratio	2.24
	Liveability	94.8%
	Shell colour	Dark brown
	End of lay body weight	2100 g

Source: Hisex Ranger Performance Data Sheet, May 1996.

Most of the modern brown egg hybrid layers will meet the requirements of the free-range producer, and it is often personal preference and circumstance which determine what is ultimately selected. Contact the various breeding companies and ask them to send details of free-range egg production and egg weight profiles for their breeds. Table 3.5 shows the average production rates for eight flocks of free-range ISA Browns in Norfolk.

The statistical information provided by breeders may include a laying graph such as the one in Figure 3.2 for the Lohmann Brown. Here are a few pointers on how to interpret it.

According to the graph, 4,590 Lohmanns that had been hatched on 13 February 1994

Hisex Brown hybrids.

Table 3.5   Hen-housed production of eight flocks of ISA Browns

Flock	Number of eggs	Number of weeks
1	320	76
2	338	78
3	327	75
4	336	76
5	329	76
6	319	75
7	326.1	75
8	329.6	74

Source: J. Vergeson's unit, Norfolk, ISA Brown Newsletter, May 1996.

Lohmann Brown - Laying Graph

Owner: BRAMLEY Farm: House: ONE Birds Housed: 4590
Hatch Date: 13/ 2/94 Date Housed: 19/ 6/94 Rearer: JEWSON Feed: DALGETY

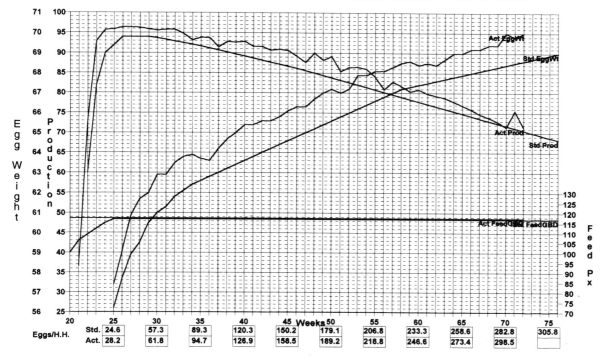

Eggs/H.H.		20	25	30	35	40	45	50	55	60	65	70	75
	Std.		24.6	57.3	89.3	120.3	150.2	179.1	206.8	233.3	258.6	282.8	305.8
	Act.		28.2	61.8	94.7	126.9	158.5	189.2	218.8	246.6	273.4	298.5	

Figure 3.2   Laying graph for the Lohmann Brown. *(Lohmann Brown & Ross Poultry)*

were housed at 20 weeks. There are two lines on the graph for each of the following factors: production or number of eggs, egg weight and feed consumed. The bottom line, in each case, is the expected or *standard result*. This flock has obviously exceeded the standard, as indicated by the *actual result*.

They commenced laying at 21 weeks. This was slightly earlier than expected, although initial feed consumption until 25 weeks was higher. Overall feed consumption was only fractionally higher than expected, at around 116g per bird per day (GBD). At 70 weeks, the actual hen-housed average was 298.5, an increase over the expected 281.8.

Breeds for the table

Traditionally, the Light Sussex was the favourite breed for table production, although many of the old heavy breeds such as the Dorking, Indian (Cornish) Game, Barred Plymouth Rock and their various crosses have been used. A particularly popular cross at one time was the Light Sussex × Indian Game, using either a Light Sussex male on Indian Game females, or vice versa. The Indian Game has a particularly broad breast, a factor which is carried through to the crossed progeny. Another favourite cross was the Rhode Island Red

(text continues on page 34)

A SELECTION OF PURE BREEDS

PHOTOGRAPHS ON PAGES 30–33 BY JOHN TARREN

Rhode Island Red male. This American breed has provided the genetic basis for all the commercial brown egg laying hybrids.

Light Sussex female. Once one of the most important table breeds, it is now kept mainly for show. It was crossed with the Rhode Island Red male to produce sex-linked first crosses that could be identified at hatching as male or female.

White Leghorn male, one of the most important of the Mediterranean breeds. It provided the genetic basis for all the commercial white egg laying hybrids.

Buff Cochin male. When the original Cochin was introduced in 1843, it created a sensation in Britain, fostering a huge interest in poultry. Along with the other Asiatic breeds, it introduced the genetic factor for brown egg shells.

Orpington male. The original Black Orpington was the first dual-purpose breed and also gave rise to the Australorp in Australia.

Maran female. Good strains of this breed produce dark-brown, speckled eggs. Like many of the traditional breeds, its productivity has declined.

Indian (Cornish) Game, once prized for its meaty broad breast. Now mainly kept for showing, it was crossed with the Light Sussex to produce table chickens in the past.

Croad Langshan male. This is one of the utility breeds that is included in the Rare Breeds Survival Trust's Poultry Project which aims to improve the laying qualities of traditional breeds.

A SELECTION OF PURE BREEDS (*continued*)

Dorking male. Its ancestors were originally brought to this country by the Romans. It has five toes rather than the usual four.

Barred Plymouth Rock hen. When crossed with a Rhode Island Red, she produces sex-linked chicks. This cross is the one on which the Black Rock and Bovans Nera are based.

The stately Brahma is one of the Asiatic breeds that introduced the factor for brown egg shells.

Welsummer, a light breed producing dark-brown speckled eggs. The breed's utility characteristics could be improved with more emphasis on selective breeding.

Dutch bantam. In small areas bantams may be more appropriate than large fowl, but their calls are strident.

Sebright female. This is purely a show bird and a visual delight with its beautiful markings.

Silkie female. This breed differs from the others in not having interlocking feather barbs so that the plumage is fluffy, like fine wool. Silkies make very good broody hens.

Cuckoo Pekin male. The females of this breed make good broodies, although they are kept primarily for showing.

Bovans Nera, a hybridised strain based on the Rhode Island Red crossed with a Barred Plymouth Rock. It has been developed as a hardy bird for outdoor conditions.

male with Light Sussex females referred to earlier.

In recent decades, with the divergence of selective breeding into the production of layers or broilers, the emphasis has been on breeding white-feathered, quick-growing birds, of which the Cobb is the prime example.

In the 1930s and 1940s the Barred Plymouth Rock had been genetically bred for meat; then in the 1950s a genetically dominant white male was developed. This was to meet the demand for white feathers for ease of processing. This dominant white male could be mated with any colour female and still give white broilers. Then, a specialised meat-type Cornish male was mated to a White

Rock female. Most Cornish lines had nearly 100 per cent pea combs (triple combs with ridges), but now they are between 90 to 100 per cent single combed. The significance of this is that fast growth is associated with the recessive single comb type, and selection for growth rate has changed the gene frequency. In the mid 1960s the Cobb 100, a cross-bred breeder, was introduced in the USA. Then, in the 1970s the Cobb 500 was introduced to Britain and has been the main table bird ever since.

Most of the breeding companies which produce laying strains also breed their own white-feathered broilers. These hybrids have all been developed for intensive indoor conditions, but they are quite capable of less intensive management. I regularly raised Cobbs for the table, keeping them in a poultry house with an attached yard, and allowing them to range on grass when weather conditions permitted. They grow more slowly in this way, but prove quite acceptable for household or specialised commercial production such as that of organic table birds. My Cobbs were usually females which do not grow as large as the males and are less likely to develop leg problems. Exercise and a lower protein ration also help to protect against this condition.

White-feathered hybrids were originally favoured because, after plucking, there are no dark stubs left in the carcase, and it was perceived that this was what the consumer demanded.

In recent years there has been an interesting reversal of this, as darker-feathered birds have become associated with non-intensive production. In France such birds are called by the general name Label Rouge (Red Label), a reference to the trade description under which they are sold. With this description, consumers can identify them as being more naturally reared. These colour-feathered birds are more slow growing and have been developed for non-intensive production, with most of their diet coming from grain. They have extensive conditions, with flocks being kept in total free-range conditions in forested regions.

French breeding companies have used traditional breeds to produce primary breeders that can be crossed with hybrid hens to produce red, black, grey or other colour variation of plumage. The small, prolific hens have a recessive gene so that the progeny always resemble the traditional male in colour. They are plump and less rangy than traditional breeds, with shorter legs and more breast meat. Sasso 551, for example, is used to produce Poulet Gaulois, while Hubbard-ISA Redbro is used for Poulet Bronze. In the USA, they are using hybrid crosses based on the New Hampshire Red to produce red-feathered table birds, and crosses based on the Australorp for black-feathered ones.

Traditional breeds and show birds

Birds kept for show must be recognised pure breeds. There is a different emphasis on what is approved of, with appearance and carriage taking precedence over utility characteristics. Each recognised breed has its own list of standards in relation to size, colour and pattern of plumage, type of comb and so on. In Britain, the individual breed clubs, under the auspices of the Poultry Club of Great Britain, organise the standards. A number of points are allocated to each feature. For example, forty-five points might be allocated for the features of the head and face, while twenty-five could be given to the colouring of the feathers. Individual breeds may have different numbers of points allocated to specific areas, but the total must always be 100. When a bird is judged at a show, the aim is to be awarded the highest possible number of points for each attribute.

When a particular breed is decided on, the best starting point is to join the breed society and obtain a copy of the appropriate standards.

There are specialist collections which are open to the public, and where birds can be viewed. Stock is also offered for sale, although some breeds are in short supply and there may be a waiting list. Examples are the Wernlas Collection in Shropshire, the Domestic Fowl Trust in Worcestershire, and the South of England Rare Breeds Centre near Ashford in Kent. Poultry shows and agricultural shows with poultry displays are also useful venues for seeing a number of breeds.

A wide variety of pure breeds are in existence, although some are now so rare as to border on extinction. Rare breeds may be considered an irrelevance by some people, but it should be remembered that they represent an invaluable genetic pool, quite apart from their intrinsic value as individual and unique species. Those who are interested in this aspect of poultry keeping should contact the Rare Poultry Society which looks after the interests of minority and rare breeds, including the autosexing breeds.

As mentioned in the previous chapter, the Rare Breeds Survival Trust has introduced a Poultry Project which seeks to improve the numbers and quality of some of the traditional utility breeds in Britain. The breeds are Croad Langshan, Dorking, Sussex, Derbyshire Redcap, Old English Pheasant Fowl, Scots Grey, Scots Dumpy and Indian (Cornish) Game. There are other birds which have been kept as utility layers in the past and which are not represented in RBST's project. These days, most are kept chiefly as show birds rather than for their productive qualities, but programmes such as this may succeed in improving their production.

Variation

Within many of the breeds listed in Table 3.6 there is a considerable variation in the type of skin, comb, feathers and colouring. There are also different varieties of the same breed. The Wyandotte, for example, has the following varieties: Barred, Black, Blue and Blue Laced, Buff and Buff Laced, Columbian, Gold Laced, Partridge, and Red and White.

Skin colour in chickens can be white, as in the Orpington and Light Sussex, or yellow as in the Brahma and Indian (Cornish) Game. Most breeds have four toes, while the Dorking, Houdan and Faverolles are blessed with five.

The type of plumage is different, according to the breed. The feathers of most breeds have tiny interlocking barbs so that the plumage has an overall smooth, silky texture. That of game birds is close and tight, hence the reference to 'hard-feathering', while all other breeds have 'soft feathering'. Silkies do not have interlocking feather barbs, so their feathers are fluffy, almost like fine wool. The Frizzle and Naked Neck are both well named, the former with its curled feathers, and the latter with a lack of feathering on its neck.

Feather patterning also shows wide variation, with barring, spangling, pencilling, and lacing. The little Silver Laced Sebright is one of the most attractive in this respect, although its fertility has declined drastically in recent years. The Speckled Sussex has beautiful speckled markings on its reddish brown plumage, with a soft grey under-feathering.

Combs vary considerably, depending on the breed. They may be single, cup, leaf, horn, rose, mulberry or pea. (The terms are good descriptions of the appearance.) A single comb may be small, medium or large, semi-erect or folded. An extension of the comb is referred to as a leader. All these characteristics are referred to in the standards for the various breeds.

With such an enormous variety of attributes, sizes, shapes, colours and patterns in the world of poultry, it is certainly true that there is something for everyone.

Speckled Sussex (left), a traditional old breed with beautiful feathering. A closer look (below) at its plumage.

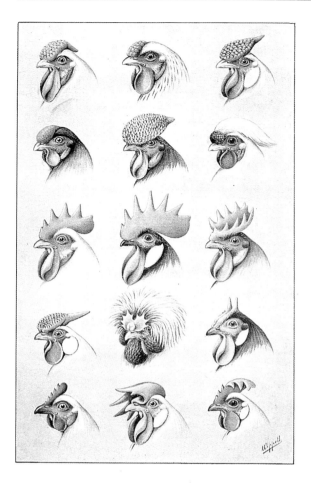

Figure 3.3 The range of combs in the domestic chicken. *(After Whippell)*

Choosing a breed

Some breeds, such as Silkies and Pekins, are admired for their brooding capacity and ability to hatch the eggs of other breeds. Poultry keepers who like the unusual may opt for breeds such as the Araucana with its greenish eggs, although the numbers of these are likely to be low. Nevertheless, selling 'multi-coloured' packs of eggs has proved to be popular. Depending on the breed and

strain of chicken, other colours include white (Leghorn), pinkish-brown (Rhode Island Red) and chocolate brown (Maran, Welsummer).

If the aim is just to have a few chickens in the garden, to supply eggs for the family, it does not matter a great deal which breed is chosen. Choose which breed you like. The old breeds will cost more to feed and they will lay fewer eggs, particularly over the winter period when they may stop laying completely. If space is really confined, consider keeping a few bantams. Some people keep chickens just because they look pretty, and what could be a better reason?

Future breed developments

In the last edition I predicted the following (The outcome is shown in brackets):
- More hybrids developed for extensive conditions. (This has happened with laying and table birds.)
- The increasing use of hybrid strains to improve the productivity of traditional pure breeds, with more emphasis on selective breeding for utility purposes. (No production statistics are yet available).
- Hybrids developed for greater docility and the banning of beak trimming in most outdoor flocks. (Beak trimming has been restricted but not banned, except in organic flocks. It is interesting to note the comment of Professor R. Pressinger: 'If future stocks with a low propensity for feather pecking, which are currently being developed, are housed in well-designed and properly managed systems then farmers will be able to control feather pecking without the need for beak trimming.'[1])
- Genetic engineering to produce transgenic chickens. [2] This is where a gene perceived to be useful is transferred from one line to another. (This is now happening in breeding research laboratories all over the world. One of the most unpleasant developments has been the breeding of featherless chickens for the intensive broiler industry. [3])

Table 3.6 Pure breeds of large fowl and bantams

Breed	Type	Origin	Eggs
Ancona	Light	Italy	White/Cream
Andalusian	Light	Spain	White
Appenzeller	Light	Switzerland	White
Araucana	Light	Chile	Blue-green
Asil (Aseel)	Heavy	India	White
Augsburger	Light	Germany	White
Australorp	Heavy	Australia	Brown
Barnevelder	Heavy	Netherlands	Brown
Belgian Game	Game	Belgium	Tinted
Belgian Barbu d'Anvers	Bantam	Belgium	White
Belgian Barbu de Watermael	Bantam	Belgium	White
Belgian Barbu D'Uccle	Bantam	Belgium	White
Booted Bantam	Bantam	Belgium/Holland	Tinted
Belgian Rumpless D'Anvers	Bantam	Belgium	White
Brabanter	Light	Netherlands	White
Brahma	Heavy	India	Tinted/Brown
Brakel	Light	Belgium	White
Breda	Light	Netherlands	White/Cream
Bresse	Heavy	France	White
Brockbar	Autosexing/Heavy	UK	Tinted/Brown
Brussbar	Autosexing/Heavy	UK	Tinted/Brown
Cambar	Autosexing/Light	UK	White
Campine	Light	Belgium	White
Carlisle Old English Game	Game	UK	Tinted
Cochin	Heavy	China	Tinted/Brown
Crève-Coeur	Heavy	France	White
Croad Langshan	Heavy/Dual	China	Brown
Dominique	Heavy	USA	Brown
Dorbar	Autosexing/Heavy	UK	Tinted
Dorking	Heavy	UK	Tinted
Dutch Bantam	Bantam	Netherlands	Tinted
Faverolles	Heavy	France	Cream
Fayoumi	Light	Egypt	Cream
Friesian	Light	Netherlands	White
Frizzle	Heavy	Asia	White/Tinted
German Langshan	Heavy	Germany	Brown
Hamburgh	Light	UK/Netherlands	White
Houdan	Light	France	White
Indian (Cornish) Game	Heavy/Game	UK	Tinted
Italiener	Light	Germany	White
Ixworth	Heavy	UK	Tinted
Japanese Bantam	Bantam	Japan	White/Cream
Jersey Giant	Heavy	USA	Tinted/Brown
Ko-Shamo Bantam	Bantam/Game	Japan	Cream
Kraienköppe	Light	Germany	White
La Flèche	Heavy	France	White/Tinted
Lakenvelder	Light	Germany	White/Tinted
Legbar (Cream)	Autosexing/Light	UK	Blue-green
Leghorn	Light	Italy	White
Malay	Game	Asia	Tinted
Marans	Heavy/Dual	France	Dark brown, speckled
Marsh Daisy	Light	UK	Tinted
Minorca	Light	Spain/Italy	White
Modern Game	Game	UK	Tinted
Modern Langshan	Heavy	Asia	Brown
Nankin Bantam	Bantam	Asia	Tinted
Nankin-Shamo Bantam	Bantam/Game	Japan	Cream
New Hampshire Red	Heavy	USA	Tinted/Brown
Norfolk Grey	Heavy/Dual	UK	Tinted
North Holland Blue	Heavy/Dual	Netherlands	Brown
Old English Pheasant Fowl	Light	UK	White
Orloff	Heavy	Russia	Tinted
Orpington	Heavy	UK	Brown
Oxford Old English Game	Game	UK	Tinted
Pekin Bantam	Bantam	Asia	Cream
Plymouth Rock	Heavy/Dual	USA	Tinted/Brown
Poland	Light	Poland	White
Redcap	Light	UK	White
Rhienlander	Light	Germany	White
Rhodebar	Autosexing/Heavy	UK	Brown
Rhode Island Red	Heavy/Dual	USA	Tinted/Brown
Rosecomb Bantam	Bantam	UK	Cream
Rumpless Araucana	Light	Chile	Blue-green
Rumpless Game	Game	UK	Tinted
Scots Dumpy	Light	UK	Tinted
Scots Grey	Light	UK	White
Sebright Bantam	Bantam	UK	Cream
Shamo	Game	Asia	Tinted
Sicilian Buttercup	Light	Italy	White
Silkie	Light	China	Cream/White
Spanish	Light	Spain	White
Sulmtaler	Heavy	Austria	Cream/Tinted
Sultan	Heavy	Turkey	White
Sumatra	Light	Asia	White
Sussex	Heavy	UK	Tinted
Transylvanian Naked Neck	Heavy	Hungary/Malta	White
Tuzo Bantam	Game/Bantam	Japan	Tinted
Uilebaard	Light	Netherlands	Tinted
Vorverk	Light	Germany	Cream/Tinted
Welbar	Autosexing/Light	UK	Dark brown
Welsummer	Light	Netherlands	Brown, speckled
Wyandotte	Heavy	USA	Tinted
Wybar	Heavy	UK	Tinted
Yamato-Gunkei	Game	Japan	Cream/Tinted
Yokohama	Light	Japan	Tinted

There are also miniature versions of many of the breeds.

These are also called bantams, although they are not true bantams.

References

1. Workshop at Department of Clinical Veterinary Science, Bristol University. June 2000.
2. ISA Brown Biotechnology and Genetic Selection in Poultry, Technical Topics No. 10, 1996.
3. Department of Agriculture. University of Tel Aviv, 2002.

CHAPTER 4 **Housing the Small Flock**

A maximum of comfort with a minimum of risk insures healthy poultry.

M.Roberts Conover, 1912

Housing must not only give shelter from the cold, rain, wind and occasionally fierce sunlight, but it must also provide adequate ventilation for the birds' well-being. It should invite the hens to lay their eggs in the nest boxes provided, as well as to feed and sleep in comfort and security. These basic principles apply to any poultry housing, large or small, commercial or domestic.

See also page 65 for the RSPCA's Freedom Food housing requirements and page 69 for Soil Association standards.

The size of house depends on the number of birds likely to be kept. Most manufacturers offer houses of varying sizes to cater for 3–4, 6–8, 12–15 birds, etc, with the lower number in each case applying to bantams, and the higher referring to large fowl. Bear in mind, however, that the latter usually means a hybrid layer. A large Brahma would be equivalent to at least two of these!

Houses are also available for the small commercial flock of 50, 100 or 300 birds. Static houses such as those used by large commercial enterprises are covered in the next chapter.

The following are the important factors to take into consideration when purchasing a house.

Timber construction

Support timbers need to be strong enough to provide a secure framework, and also able to stand up to the rigours of moving the house.

They should be at least 2.5 cm thick, and ideally 3 cm. Screws, hinges and other metal attachments should be galvanised so that they do not rust.

All the wood must be proofed against damp or it will soon begin to rot. The options are tanalised or protimised timber; these are preferable to the more toxic creosote. Painting on proofing is not as effective as soaking timbers in it before construction. The best method is to force the proofer under pressure, deep into the wood. An example is the tanalith process, which lasts for many years.

Walls

The cladding of the walls needs to be substantial enough to provide good insulation as well as weather protection. Depending on the manufacturer, the walls may be exterior-grade plywood, matchboarding or tongue-and-groove boards. Whatever the construction, the wall needs to repel rainwater and be proofed against creeping damp. Patches of damp appearing on the inside are sure indications that the cladding is inferior and needs urgent attention. Depending on the situation, this will be either recourse to the supplier or applying an extra coat of proofing.

In exposed areas it is more appropriate to have overlapping weatherboarding with internal insulation. I well remember having a letter from a poultry keeper in Scotland some years ago. One bitter winter's morning she found her chickens dead, frozen to their perches. Most poultry books, she told me, were written by people who live in mild areas, and who have no idea of what really severe winters entail. Some time later, I visited Finland and found that there, all free-range poultry are housed in winter. The floors have a thick layer of wood shavings which is added to throughout the winter. The chickens are kept in buildings with triple insulation, heated by stoves fuelled by the never-ending supply of wood from the plentiful birch forests. Without these conditions, they would perish.

In really exposed areas of Britain, it is much better to follow the Finnish example and adapt a barn or shed for winter occupation, although wood stoves would be an unnecessary luxury. Free-ranging is not always a sensible option. There are times when barn eggs are more appropriate.

If you need to add insulation yourself, insulation boards are widely available in DIY stores. Take care not to obstruct the air inlets of the house.

See also section on insulation in the next chapter.

Particularly windy places, such as coastal areas, may require the provision of storm supports for a small house. A simple way of providing this in winter is to pass a strong rope over each end of the house and peg it securely into the ground.

Roof

The roof is obviously an important area, not only to keep out the weather, but to retain warmth. It should also provide an overhang for shedding water and providing shelter immediately outside the house. It may be made of shiplap boarding, wood covered with bitumenised or tarred felt, or an Onduline corrugated bitumen. Where bitumenised felt is used, make sure that it covers the whole roof and overhang, without gaps. It should be securely attached, either to the rim of the overhang, or underneath it.

Whatever the type and construction of the roof, it needs to be sturdy and totally waterproof.

Adequate ventilation is vital as moist, stuffy conditions often lead to health problems such as respiratory complaints. The aim is to provide fresh air while excluding draughts. Larger moveable houses have ridge ventilation. There is a gap at the apex of the roof, protected by a ridge, so that air can pass through, while rain is kept out. The air enters from side eaves or apertures, which are often protected by a baffle board. As air becomes warm, it rises to go out at the top. Larger houses often have windows to provide light. Polycarbonate is sometimes used in preference to glass because it has better insulation properties and is less likely to break.

Smaller houses usually have a window

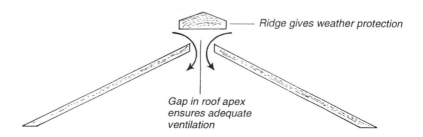

Figure 4.1   Ridge ventilation.

vent covered with galvanised steel mesh. This is placed high up on one wall, above the heads of the perching chickens. Some houses may also have a sliding partition or window shutter for severe weather protection. The house should always be sited so that the window is on the leeward side.

Floor

Most smaller houses come complete with a solid wooden floor with supports that keep it clear of the ground. As most droppings tend to accumulate under the perch, it is a good idea to have either a droppings board or a box underneath it, which can be cleared periodically. A droppings board is a section of wood that slides out for easy removal of droppings. Strong polythene sheeting is a good alternative.

A slatted floor is often found in larger houses, but is also available for small ones. This usually has 3 cm wide wooden or plastic slats placed 2.5 cm apart so that droppings can fall through to the ground below. Alternatively, a wooden-framed floor with mesh panels can be used. The advantage of a slatted floor is that droppings fall through and when the house is moved, the droppings can then be dispersed.

Perches

Every house should come with one or more perches, depending on the number of birds for which it was designed. The perch should be placed higher than the nest boxes so that the chickens are not encouraged to sleep in them.

The ideal perch is rounded off and smooth, to provide a comfortable surface for the foot to grasp. The width is around 4 to 5 cm. (If you have very large fowl, the width may need to be slightly greater.)

A depth of the same size is adequate, allowing the perch to fit into prepared wall sockets at each end. These should be made in such a way that the perches can be easily

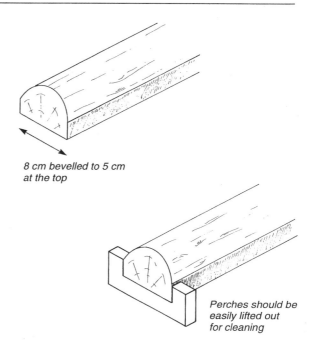

8 cm bevelled to 5 cm at the top

Perches should be easily lifted out for cleaning

Figure 4.2 Single perch.

lifted out for cleaning because they are one of the favourite hiding places of the red mite, a tiny, blood-sucking parasite which comes out at night when the chickens are perching.

The perch should not be more than 60cm above the floor for heavy birds; otherwise damage to the feet may result from continual jumping down. Small abrasions can lead to infections and abscesses of the feet, a condition commonly referred to as bumblefoot. Details of this are given in Chapter 13. For light birds or egg laying hybrids, the distance can be greater, but research at the Scottish Agricultural Colleges suggests that this should not exceed 1 metre.[1]

Where several perches are used, the distance between them needs to be around 30 to 40 cm, depending on whether they are parallel or stepped. With stepped perches, the height above the floor will vary, but the bottom one should not be more than 60 cm above the nearest surface. Where single

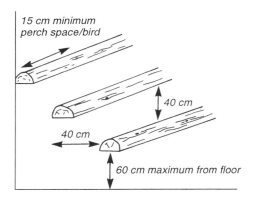

Figure 4.3   Stepped perches.

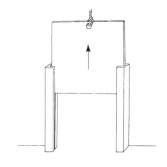

Figure 4.4   Lift-up pop-hole entrance.

perches are used with a droppings board, the height is normally about 20 cm above the board.

Pop-holes

A pop-hole is a small entrance/exit for the chickens. It is quite separate from the main door of the house, which is there for the poultry keeper's use.

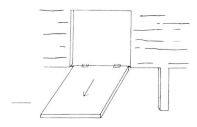

Figure 4.5   Drop-down ramp pop-hole.

The principle of the pop-hole is that it can be opened from the outside and used to control access to a given area. In a house with a raised floor, the pop-hole will need an exit ramp so that the birds do not have to jump down or flap their way upwards. The simplest way of providing this is to have a hinged door which drops down from the top and becomes the ramp. Alternatively, the ramp can be a permanent fixture, with a door which slides up and down for opening and closing. If there is an attached run, it is useful to be able to open or close the pop-hole door from outside; otherwise you will need to go into the run. All doors, pop-holes and windows should fasten firmly to exclude predators.

Constructing a porch protection for the

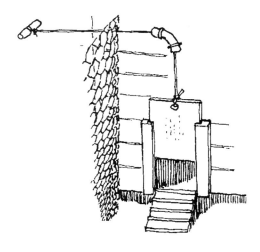

Figure 4.6   A simple way of opening a pop-hole without going into the run.

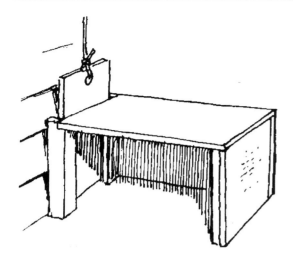

Figure 4.7 A porch placed in front of the pop-hole prevents draughts.

Figure 4.8
Static box fitted with plastic curtain to discourage egg eating.

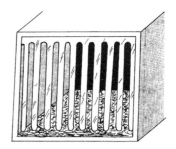

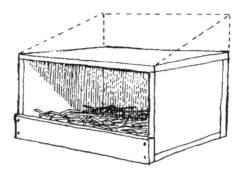

Figure 4.9 Nest boxes inside a house will benefit from having a sloped roof to prevent birds using them as perches.

pop-hole, as shown above in Figure 4.7, is an effective way of cutting out draughts in the house.

Nest boxes

In a small house, nest boxes need to be placed lower than the perches and in the darkest area of the house because this will not only attract the chickens to go in and lay their eggs, but will also discourage egg eating. They should be above ground level, however, so that floor-laid eggs are not encouraged. Ideally, they should be accessible to the poultry keeper from outside the house. This involves having a hinged, waterproof lid above the nesting area. As a general rule, traditional nest boxes, such as the one shown in Figure 4.9, are around 30 cm high × 30 cm deep × 25 cm high. A panel 8 cm wide along the bottom of the entrance will ensure that the nesting material does not fall out. Some nest boxes are made with wire mesh bases which help to keep them cool and avoid triggering broodiness.

Large breeds may need larger nest boxes.

Where free-standing nest boxes are used in a building such as a barn or shed, it is useful to put a section of sloping hardboard along the top. This will prevent birds from perching on top of the nest boxes and fouling the area around them.

Where a bank of nest boxes is raised off the floor, there should be alighting perches in front so that entry is facilitated.

A small domestic house for 6 to 12 birds functions best with one nest box to every 3 birds. Larger houses need one for every 5 birds.

Wood shavings make a good nesting material, as long as they are clean, dry, free of dust and non-toxic. In other words, they should be from timber that has not been treated chemically, as poultry have been poisoned in this way. Wood shavings should come from specialist suppliers.

Synthetic materials are also used in nest boxes. For example, there is a plastic material which is formed into tussocks resembling grass. Any droppings tend to fall between the projections, leaving the eggs unaffected. The material is available in rolls and can be cut to fit the nest box. To clean it, remove the square and shake it so that the dried droppings fall out. It is easily washed in hot water.

It is worth remembering, however, that hens prefer nest boxes containing loose material which can be moulded into shape by the body. When given a choice in one trial, they invariably chose the nest box with loose material. It is always satisfying when scientific research confirms what has always been obvious.

Straw is often used, but it should be free of mites. Hay should not be used because, if it gets damp, it can quickly become mouldy, leading to respiratory complaints in both birds and poultry keeper.

Doors

There should be a door for the poultry keeper. This is usually a normal size door in a larger house, or a lift-up section of the roof in a smaller one. Whatever it is, it needs to provide easy and covenient access to the inside of the house.

Security

Where there is a door, there should be a well-oiled bolt to make sure that it stays shut at night. It may also be worth putting on a padlock. There have been many incidences of theft, particularly of rare breeds. Some poultry keepers have even had to install an alarm system. A well-trained watchdog is a positive asset in these circumstances. One lady bought some geese to act as watchdogs.

Their clarion calls and hisses were most effective.

Ease of movement

It is important that a house which needs to be moved regularly is equipped for ease of movement. Depending on the size, it will have wheels or skids, with handles or a towbar. Larger houses can be hooked to a car or tractor to be dragged. Smaller houses usually have carrying handles, enabling two people to move them. If a house is to be moved by one person, it really needs to be on wheels.

Some modern houses have been designed so that the whole structure can be dismantled and re-erected in a short time. This simplifies cleaning enormously.

Types of house

There are several different types of house, depending on the size of flock and the area of ground available. For the small poultry keeper the choice is essentially as follows.

Some small houses can be completely dismantled for cleaning. *(Forsham Cottage Arks)*

Small house with attached run. *(Littleacre Products)*

Integral house and run

This is an option for the small garden where only a few birds are kept. It is a unit made up of a house with the run extending beneath it, maximising the use of space. The chickens are able to go outside and have protection from the rain.

As the unit is moveable, it is suitable for use on a lawn, with fresh grass being made available on a regular basis. The disadvantage of this type of house is that extending it with another run may be difficult, if this should become necessary. Some manufacturers, however, do provide extension runs for their whole range.

Those with larger flocks may also find this type of house useful, particularly for breeding or rearing purposes. A male and several females of the same breed can be kept here while fertile eggs are collected for incubation. A broody hen with chicks would also find it a congenial home, as long as there is a suitable ramp for the chicks.

Small box-type and A-shaped grazing arks are available, with a shelter attached at one end. This type of structure is useful for summer grazing, or for a broody hen and chicks, but does not replace a permanent house.

Separate house and run

Most manufacturers produce a wide range of

Moveable house for the small free-range flock. The feeder on the left is protected against the weather. *(Forsham Cottage Arks)*

A moveable house for up to 100 birds (above), suitable for the small commercial or organic flock. It is on skids for moving and has nest boxes on either side.

Inside the house (right). Note the hinged perches that are put up during the day.
(Gardencraft)

different sized houses, catering for everything from a trio of chickens to around 300. Runs are also available for attaching to the end where small numbers of birds are involved. Larger houses will normally be free-standing in a paddock where there is perimeter fencing to keep out foxes. These houses can also benefit from having an attached run, particularly one which is partially roofed, because it provides a dry area immediately outside the house. If the other end of the run is open, the birds can still free-range further afield.

It obviously makes sense to buy a house and run from the same manufacturer in order to ensure that they fit together properly. Some manufacturers supply extension units to existing runs, if a longer ranging area is

A traditional layers' house refurbished for current day use. Straw is being used to prevent muddy conditions outside the house so that the hens' feet are clean when they go in. This helps to keep the eggs clean. *(Onduline)*

required. Runs are not too difficult to construct by those reasonably competent at carpentry.

A house with a run that can be separated from it is easier to move as two units. The birds can then be confined to one or the other while in transit to a new grazing area.

House with alternate runs

Some poultry keepers have two distinct areas for their birds, depending on whether it is summer or winter. The summer accommodation is a small, moveable house and run on grass. In winter, a garden shed or moveable house is used, with access to a concrete run. Hens cannot scratch about on concrete, but the provision of a large, shallow box of clean sand will keep them happy in the short term. Ground which is overused can become a haven for parasites and other disease organisms. Traditionally it was called 'sick' ground. It can also become a mud bath in winter. If it is not possible to provide an alternate run, it is infinitely preferable to concrete the run so that it can be easily hosed down.

A thick layer of chipped wood is effective in

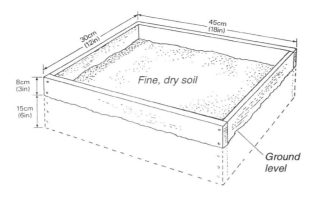

Figure 4.10 Dust bathing area for hens with small runs or on concrete.

Chipped wood being used as a ground protector and scratching area for these Cobb table birds.

a winter run. It is quick-draining and clean. Raking it periodically and topping it up keeps it in good condition right through the winter. When the birds are returned to their summer housing, the layer of wood chips makes an excellent mulch for the garden.

Making a run

If the house does not have a run, the chickens will need to be confined so that they cannot escape to wander on other people's land, and they are protected against predators. On a

A small run made of galvanised metal tubes with garden netting on the top and sides. *(Biodesign)*

small scale, weather-proofed posts can be erected with galvanised poultry wire netting in between. Green plastic coated netting is also available, and is slightly more decorative for garden use. This will need to be pegged into the ground or the birds will escape underneath it. Poultry netting also has a tendency to sag, so will need to be strained, as shown in the illustration. It will not generally keep out dogs, foxes and cats, but in a protected garden, these may not be a problem.

Where top protection is needed, plastic anti-bird netting can be used. Designed for keeping wild birds out of fruit cages, it will deter cats and dogs from going in, as long as it is pulled taut across the top. It is also effective in providing a degree of shade cover in summer. Heavy-duty plastic mesh can be used to provide windbreaks.

Fencing on a larger scale is covered in Chapter 6.

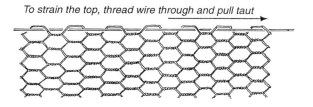

To strain the top, thread wire through and pull taut

Figure 4.11   If poultry netting is used to make a run, it needs to be pulled taut at the top.

Fittings

Every house needs to be furnished with the appropriate fittings to make it a habitable domain for the birds. These are in addition to the perch, nest boxes and pop-holes which are normally provided when a house is bought.

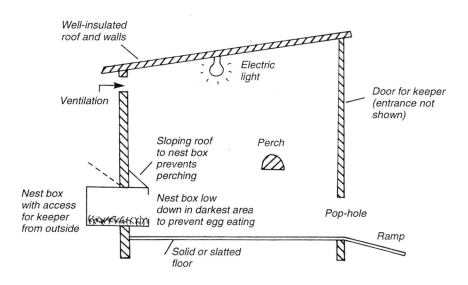

Figure 4.12   Interior of a small perchery house.

Feeders

Depending on the size and design of the house, the feeder may be in the house or placed in the run. If it is in the run, cover it against the weather.

Feeders are generally of two types – ones that stand on the ground, and those that are suspended. It does not matter which is used, as long as the feed is positioned high enough so that it cannot be fouled by droppings. The suspended ones are popular with many poultry keepers because they are easier to keep clean. They are made of plastic or galvanised steel, and are available in various sizes, depending on the age of the birds. Those standing on the ground must be heavy enough to avoid being knocked over.

Feeders are normally used for layer's mash or pellets, while grain is often scattered on the ground to allow the birds to scratch for it.

There are some feeders that dispense feed or grain when a chicken displaces a small bar underneath the container. Originally designed for pheasants, these have proved useful in encouraging chickens to range further afield from the area immediately around the house. Some of these feeders can also be programmed to dispense feed at certain times of the day.

Drinkers

Most drinkers operate on a gravity principle; as water is drunk from the bottom, it is replaced from a reservoir tank, which is covered by a lid to prevent contamination. Suspended drinkers which are filled manually are normally used with small houses. They can be made of plastic or

Moveable house for up to 130 birds. It has draw-out droppings boards. Outside is a demand feeder which releases food when the lever underneath it is displaced by a chicken. *(Domestic Fowl Trust)*

galvanised metal. There are also drinkers which stand on the ground. They should be placed so that litter and droppings do not find their way into the water. Depending on the unit, a drinker may be in the house or in the run. If outside, shade and rain protection are a good idea.

Drinkers need to be cleaned and filled at least once a day. In hot weather it may be more frequent. Five hybrid hens will normally drink 1 litre of water a day, but in hot weather, twice this much.

Feed storage

All feedstuffs should be stored in dry, rodent-proof containers. Purpose-made feed bins with tightly fitting lids are available. An alternative is dustbins. These provide excellent, low cost storage, and the only other requirement is a good scoop for extracting the feed.

Lighting

The provision of a little artificial light in winter will ensure that the hens continue to lay. For a small house, a 12 volt battery and bulb will suffice, although a timer will be needed unless it is switched on and off manually. Further details are given in the next chapter.

Routines

There are many possible routines for looking after a small number of chickens. People who go out to work during the day will have to time their poultry activities for the morning and late afternoon. This is not too difficult to arrange, as long as the chickens are in a protected area with shelter and shade, and the feeder and drinker have been filled. Things are more difficult during holiday periods, when arrangements must be made

for a relative, friend or neighbour to come in and feed them. This is not usually a problem, particularly if the helper is allowed to take the eggs as a reward.

Daily

- Check that the run is closed properly, then open the poultry house pop-hole. Good morning, chickens!
- Check that each bird is all right (no limps, abnormal behaviour or symptoms).
- Provide compound feed (mash powder or pellets) in a clean feeder.
- Check that the drinker is clean and fill it with fresh water.
- Check that the chickens have grit and crushed oystershell.
- Collect eggs.
- Check nest boxes and remove soiled lining material; replace with fresh.
- Give grain as a scratch feed in the afternoon.
- Spend some time each day talking to the chickens; they are full of interest.
- Check that all the birds are inside the house and close the pop-hole. Good night, chickens!

Weekly

- Remove droppings board or liner and add the droppings to the compost heap. (Do not let the chickens have access to the compost heap.)
- Clean the board or replace liner.
- Brush out the house if necessary.
- Move the house and run to a new area of grass.

Further details of these activities are in the Poultry Management chapter.

References

1. Scottish Agricultural Colleges (SAC), 1995.

CHAPTER 5 Housing the Large Flock

The fears of planners and objectors are not necessarily rational, but you must not dismiss them, or those who raise them.

Ray Williams
ADAS Land Management & Utilities Team

Those with large commercial flocks often use a static house in an area fenced off by a permanent perimeter fence. Within this area, the ground may be divided into paddocks which are made available in rotation. Portable electric netting is used to control paddock access. (See Chapter 6 on Land Management.)

I must admit that I do not favour massive houses. The flocks are so large that it is inevitably difficult to spot incipient problems and potential illness. It is also far more difficult to organise rotational grazing and keep the air in the house clean and free of dust and ammonia. Smaller static houses are much easier to manage, although the ground area immediately outside all houses which remain on the same spot will become overused and bare. I have yet to see a static free-range unit where this problem has been effectively solved.

Planning

Planning permission is not required to house and keep a small domestic flock as long as the house is a small, moveable structure. For larger flocks and houses, the situation is more complex. First of all, it is vital to try to establish whether planning permission is likely to be agreed before spending a lot of money on setting-up costs. Secondly, it is a good idea to find out what criteria are considered important by planners, and to make sure that there is nothing in the enterprise that would fail to meet these requirements. Thirdly, clear information is essential, for planning authorities are not necessarily well informed on agricultural matters. For example, 'factory farming' houses are generally unpopular, and it is not unknown for planners to have overlooked the fact that a large house is a free-range one. Make sure they know that yours is not a battery cage house!

Environmental impact is an important consideration. Houses need to blend into the landscape rather than dominate it. The general design and colour of the building should be sympathetic rather than strident. Siting is a vital consideration, particularly in National Parks and Areas of Outstanding Natural Beauty. A certain amount of landscaping may be necessary in order to conceal obtrusive elements such as feed bins.

The impact of traffic is considered in relation to the existing road system. If it is not capable of supporting more vehicles, the applicant may be required to provide lay-bys and turning spaces.

Other concerns are manure disposal, odours, dust, rodents and noise. Any application needs to have considered these factors thoroughly and to provide evidence that there will be no problems in these areas.

ADAS offers an Environmental Impact Assessment service in this respect, as well as producing a commissioned report on individual sites for the planning inspectorate.

I am grateful to planning expert Gordon Holt for providing the following detailed information on planning requirements.

Most poultry houses are 'development' needing planning permission, but in some instances a general permission is given by planning law, and buildings are deemed 'permitted development', subject to a 'notification' procedure. This means that local authorities have to be informed of the intention to erect a farm building, and if they feel it would have a 'significant effect' on its surroundings, full details have to be submitted. If 28 days elapse after notification and nothing is heard from a local authority, 'permission' may be assumed. However, if particulars of the building are required, the principle of the building may not be challenged – just matters of siting and design.

Holdings over 5 hectares

If the following criteria are complied with, the notification procedure mentioned above applies.

- The poultry buildings must be designed for agricultural purposes (i.e. not an old prefabricated building initially designed for another purpose).
- There must be no dwellings or workplace on nearby land not part of the holding, the boundary of which is nearer than 400 metres to the proposed building.
- There must be no classified road nearer than 25 metres.
- The building must not be larger than 465 square metres.
- The height of the building including any hoppers must not exceed 12 metres.

If a building does not comply with these rules, planning permission is required in the normal way.

Holdings over 0.4 hectare (1 acre) and under 5 hectares

Conventional planning permission is required for all free-standing poultry houses. In other cases such as where an existing building is being extended by up to 10 per cent, and the above conditions complied with, such developments are 'permitted development'. However, in National Parks and the Broads, the notification process comes into play.

Holdings of less than 0.4 hectare (1 acre)

Planning permission is required for any permanent poultry building.

Moveable buildings

If a poultry shelter or hut is moveable, it is possible to argue that no 'development' has taken place, and that the structure is what lawyers call a 'chattel'. However, cases have shown that chattel status is not easy to achieve. For instance, in a Scottish case (Cunninghame District Council, 12 January 1991) a wooden free-range poultry building measuring 3.7 metres by 2.3 metres was held too substantial to be termed portable and it had to be 'erected'. In another case (Three Rivers District Council, 12 September 1989) chicken houses measuring 2.45 metres by 2.5 metres which just rested on the ground were considered to have a degree of permanence given by their incorporation into a structure of fenced-off runs. In addition the huts did not appear to be self-contained structures brought onto the land in one piece.

As may be detected in these cases, in order not to be 'development', a small poultry building would need to be able to be brought to a site in one piece, and once there would need to be readily capable of being moved around a field without needing to be dismantled. In addition any feature such as a concrete base or physical attachment to the ground would give the building a degree of permanence and bring it within planning control.

Planning issues

The main issues that are raised when planning applications for poultry houses are submitted to local authorities are:

- whether there would be harm to the rural character and appearance of the area
- whether any additional traffic generated would create highway safety problems or affect the amenity of others
- Whether there would be any significant problems for neighbours deriving from noise or smell

The viability of a holding is not normally a consideration which needs to be taken into account. General Government advice is that planning decisions should not be based on considerations of whether businesses would be commercial successes or not. However, if it can be shown that an enterprise is of doubtful viability and the results of failure would leave a building in the countryside which would remain unused, or could then justifiably be reutilised for a less suitable purpose, viability may be brought into play as an issue.

On the other hand, where a free-range operation is already running and additional buildings are required to increase production or efficiency, agricultural need may be given some weight, but as a rule it does not over-ride other significant objections.

Obviously planning problems are magnified when poultry houses are required for intensive rearing, but the following appeal case decisions demonstrate issues that have arisen in connection with housing for free-range birds.

* A 4,000 hen free-range egg production unit was proposed. A long accessway was needed and although an inspector felt that removal of a length of roadside hedgerow for visibility would be acceptable, he considered that a scar across an open field would be created which would have an adverse effect on the area's character and amenity.

Objections were received from nearby residents that there would be smell and vermin problems, but the inspector noted evidence that because of the high volume of air exchange in relation to relatively low stocking rates, very little smell would be produced. The premises would be subject to regular inspection and the design of the building facilitated easy control of vermin. The appeal was dismissed because of the access objection. (Monmouth District Council, 5 August 1993).

* Permission was requested for 9 permanent ranger houses, each housing 300 birds, and for 10 mobile houses for meat birds, each housing 100 birds. It was intended to graze the land, which was in a green belt area, with cattle alternating with the birds. Although the council questioned the viability of the project, an inspector was satisfied that this was not an issue. He considered objections that noise and smell would affect nearby houses but felt that this would not be a problem. The appeal was allowed subject to there being no outside storage or burning of waste, and the approval of a scheme for the disposal of all wastes from the bird houses. (Tandridge District Council, 25 January 1993.)

* An application was made for a 46 metre long polytunnel to be used in connection with free-range production. An inspector concluded that the polytunnel would be screened and would not look incongruous. Visibility would be slightly substandard at the access to the lane but given the low level of use there would be no traffic hazard. There would be no smell problem if the deep litter were removed from the polytunnel periodically. Noise from frightened or excited chickens was dismissed as an objection, particularly as other animals which

would also make a noise could be kept on the land. He noted that a flock roaming across the fields would be far less likely to create a disturbance than might be the case at more intensive operations. (Teesdale District Council, 25 June 1992.)

* A 6,500 square foot 400 free-range hen building was proposed in Wales. A local authority argued that the land was poor and wet which would lead to problems of caked litter giving rise to smells. It also asserted that the enterprise was not viable due to conditions which would give rise to poor production and high mortality from parasitic diseases. In addition, objection was made to the exposed position of the proposed building in open countryside.

An inspector felt that he had insufficient evidence to judge that the birds would not range, but he did express considerable doubt as to the viability. He mused that if the business failed, the council could be faced with an application for a more intensive livestock operation that would have serious consequences for those living nearby. On the third issue he did not see how the low profile structure would have any adverse impact on the character and appearance of the area, nor would there be any amenity problem for nearby residents from the type of operation proposed. The appeal was dismissed. (Vale of Glamorgan Borough Council, 8 May 1992.)

* An inspector reasoned that a building already erected without planning permission in a green belt area, and proposed to be used in conjunction with a 3,600 free-range bird operation otherwise using arks, should not be allowed to remain. The site was in a green belt and the viability of the proposed enterprise was so in doubt that it was likely that the building would be left empty or have to be used for some non-agricultural use, and this would be harmful to green belt policy. (Thurrock Borough Council, 26 March 1993.)

Dwellings at free-range enterprises

The same planning rules apply as with other types of agricultural dwelling. When faced with a planning application for a dwelling, which could be in the form of a house or a caravan, the first question that a planning authority will ask is whether there is a need for the farmer or worker for whom the dwelling is intended, to live on the holding itself in order to maintain 24 hour supervision. If this can be established it will then have to be shown that the free-range enterprise is genuine, realistic and sustainable. These criteria amount to what is known in planning jargon terms as the 'functional test'. Of course, if a free-range enterprise is already up and running the latter conditions are a lot easier to demonstrate than if it is in embryo form, or if it only exists on paper.

If the above criteria can be satisfied, Government advice is that detailed financial testing as to whether a holding is 'viable' is not normally necessary, and such calculations should only be brought into play in cases where the 'functional test' is inconclusive. However, in practice it is not always easy for a decision to be made as to whether an agricultural activity is genuine, realistic and sustainable without delving into accounts.

If only a caravan or mobile home is required the kind of testing described above may be a little less rigorous as local authorities know that they have the power to grant permission for a limited period after which a caravan can be required to be moved. Thus, when there is some doubt as to whether or not an enterprise would succeed in the long term, to permit a caravan for a period of two or three years provides an opportunity for a 'trial run'. At the expiry of this period the progress that has been made with the 'justifying' enterprise can be re-examined. If it appears that the hopes for a successful ongoing business are not, after all, going to be realised, permission for retention of the caravan may be refused. If there is still some doubt about the operation, but a fair chance of success given more time, a renewal of the limited period permission may be granted. In the event that the operation has proven itself, planning permission for a permanent house or bungalow is possible, as shown in the Wokingham case described below.

The following planning appeal decisions throw some light on various situations where a dwelling or mobile home has been refused planning permission by a local authority, and the matter has gone to the planning inspectorate for adjudication.

* A mobile home was required for a free-range poultry holding of 4.29 hectares where two houses of a floor area of 450 square metres had been erected. More than 4,000 birds were kept, together with 20 sheep. An ADAS appraisal indicated that the holding was a viable proposition and that a worker was needed on site. It stated that if not, the welfare of the poultry would be jeopardised and the close environmental control necessary might not be successfully monitored. An inspector accepted this and other evidence that the risk of predators was another factor supporting the need for accommodation on site. A three year limited period permission was given. (Crewe and Nantwich Borough Council, 26 July 1989.)

* A dwelling was requested at a free-range holding of 6 hectares supporting 6,000 birds in two large chicken houses. There was a caravan at the site already. An ADAS report endorsed the fact that the holding passed the 'functional test'. This report stated that a full time worker was needed on site. In summer the working day was long and the hens continued to range outside until dusk. Recommended codes of practice from MAFF required prompt action to deal with distressed and injured birds. The automatic ventilation systems were crucial to their well-being. Given the number and monetary value of the flock, remote alarm systems were inappropriate.

The inspector who took the appeal noted that there had been recent capital investment in a new chicken house costing £18,000 which indicated genuine intention and commitment. In view of these matters a financial test was not necessary. The appeal was allowed and the appellants applied for costs from the local authority. It was concluded that the council had no sound evidence to base their opposition to the proposal, and it was unreasonable to expect the appellants to continue to live in a caravan. As a result the council were required to pay the costs of the inquiry. (Horsham District Council, 2 September 1994.)

* Enforcement action was taken against an unauthorised caravan at a two acre smallholding where there were 600 free-range birds housed in three sheds. There was also a new barn egg unit on site. An inspector applied the 'functional test' and found that it was necessary for someone to live on the holding to care for the birds. Considerable attention was paid to the question of whether there was a reasonable chance of the holding becoming viable, and an ADAS business appraisal was available at the inquiry. This assumed that there would be a flock of 400 free-range birds and a barn egg unit capable of housing 1,800 hens. On this basis the holding could be viable, but it was noted that much depended on the marketing skills of the appellant.

The local authority contested the view that the holding could ever become viable on such a small site, and the inspector noted for himself that the free-range paddock was devoid of grass. He noted that MAFF standards indicated that the maximum number of hens that could be accommodated in this paddock was 490. It was concluded that the ultimate success of this enterprise could depend on an expansion of the barn egg production side, but set against this was the fact that such eggs were more difficult to market than the free-range product. The balance was tipped in favour of a temporary permission for the mobile home by the fact that there had been considerable investment (£18,000) in the barn egg unit. (Wealden District Council, 24 May 1993.)

* Permission had been refused for a permanent house at a free-range poultry holding. A limited period permission for a caravan had enabled the number of hens to the present figure of 1,200. An inspector was satisfied that there was a need for the farmer to live at the holding in order to supervise the flock and prevent vandalism.

The economics of the holding were disputed but the inspector noted ADAS and NFU advice that a free-range unit of this size could be viable. There was a ready market for free-range eggs at a premium price. The farmhouse was allowed. (Wokingham District Council, 3 October 1986.)

Occupancy conditions

A planning permission for any house or bungalow permitted in the countryside, where the justification for the dwelling is agricultural need of whatever sort, will have the standard agricultural occupancy condition imposed. This requires that the person occupying the dwelling is solely or mainly employed in agriculture (defined generally). The effect of such a condition is that if anyone who does not comply with the condition occupies the dwelling, they are acting unlawfully and a local authority could take an enforcement action to secure compliance. However, it is far commoner for an application to be made to remove the condition. Such applications may be successful if it can be shown that there is no longer any agricultural need for the dwelling at the holding in question, and also that unsuccessful attempts have been made to sell the dwelling at a price which reflects the fact of condition. Normally, the 'encumbrance' of

an agricultural occupancy condition reduces the market value of a property by about 30 per cent. However, local authorities are very reluctant to remove any agricultural occupancy conditions and resort to appeal may have to be made. Current statistics show that about 35 per cent of such appeals are successful.

In the case of caravans which may have been permitted to allow an agricultural enterprise to develop, the normal occupancy condition may be applied, but it is more likely that a limited period permission may be given for two or three years. Such a condition requires that the caravan use be ceased at the expiry of that period of time.

This type of permission may be renewed if it is felt that more time is needed to develop a holding to the point where a permanent dwelling could be allowed. However, if there is little sign of this condition coming to pass, and the caravan remains after permission has expired, enforcement may be taken.

Farm shops

The main diversification employed by free-range smallholders is to set up a small egg shop at their premises, possibly combined with a delivery round. Such a use does not require planning permission at all, if only products derived from the holding are sold. However, as soon as other goods are sold which have to be 'imported' to the shop from outside sources, a point is reached when the shop is no longer considered to be ancillary to an agricultural use. There is no rule of thumb, but when the level of imported goods rises to 10 per cent, questions will start to be asked.

The sale of poultry meat produced at the farm obviously falls under the same heading as eggs, but a problem which may occur is that the business of killing, plucking and dressing, etc, may be considered an 'industrial' process which is not ancillary to agriculture. This point was demonstrated in an appeal case (Bournemouth Borough Council, 13 July 1984) where although it was confirmed that the sale of eggs from a 5,000 bird farm did not need permission, no matter what the numbers involved, it was considered that poultry sold for eating was a 'processed' item and therefore its sale did need permission. Although it may be difficult to fathom the logic behind this thinking, which in effect means that live birds may be sold but not dead ones, the strict legal position is that an 'industrial process' has intervened between the basic farm product and what is offered for sale, thus breaking the direct linkage between farm and shop.

If planning permission is required for a farm shop this does not necessarily mean that it will be refused. The commonest reason for refusal relates to the harm which could be caused to amenity of traffic safety considerations. In addition, most local authorities are very wary that if they permit a small farm shop, it may grow into something larger which could create planning problems. In fact there is a wide range of conditions which can allow them to restrict any undesirable expansion of a farm shop.

Site

As well as meeting the planning requirements, the site should also be well drained and protected from the prevailing winds. Also essential are a mains water supply, electricity and good road access.

Although a slightly sloping site is good for drainage, the house itself will need a level site for the concrete base. Drainage pipes can be incorporated to drain water away from the building perimeter, and it is a good idea to provide an extended area of rapid drainage outside the house. This can be a wide, shallow trench filled with shingle, chalk or other suitable draining material.

The house

Fixed houses which are used for free-range production may be of the perchery or deep litter type. The framework can be timber or steel, with a concrete floor and block foundations. If an all-slatted floor is preferred, there must be sufficient depth to allow for a droppings pit underneath from which the droppings are periodically removed, normally at the end of flock lay. Wooden or plastic slats or a wire mesh floor with supporting struts are commonly used. If a wooden slat breaks, it is often possible to replace it with a less expensive plastic one. This can be attached with nails or heavy duty staples, as is the case with wood. Slats are normally 3 cm wide, spaced 2.5 cm apart. This gap between the slats is big enough to allow droppings to fall through, but not enough to trap a bird.

Many large producers prefer slatted floors because there is less likelihood of floor-laid eggs, but they have to be mindful of the welfare requirements of access to a scratching area. Some Freedom Food producers have a series of porches attached to the houses. These have pop-holes that are closed at night, but the birds still have access to the scratching area. A system such as this helps to keep eggs clean, as well as catering for one of the hen's basic instincts.

The porches are also effective in cutting out draughts. Wind funnelling can sometimes be a problem in static houses, and unlike moveable houses, they cannot be turned away from the prevailing wind. Bales of straw outside have been used to good effect as windbreaks.

If there are no porches there should be at least an overhang of the roof to provide a protected, dry area outside the walls. This can be concreted or slatted. Alternatively, a framework with mesh panels can be used as a walkway so that the hens' feet are cleaned before they enter the house. Sections of plastic-coated steel mesh are available.

Large house with an overhang to provide a protected area immediately outside the house. *(Onduline)*

Although designed as internal flooring, they make effective verandahs for the outside.

Walls are usually timber while the roof may be polyester-coated pressed steel, which is available in a number of colours. One that blends sympathetically with the environment is obviously preferable. Aluminium sheeting with internal fibreglass insulation is also common as a roofing material, as well as synthetic materials such as Onduline.

Insulation

The optimum temperature for a chicken is 21°C. At lower temperatures it will consume more food in order to keep itself warm.

The house pop-holes on the left open onto a protected scratching area which has its own pop-holes on the right.

Insulating a house will not only provide a more congenial environment for the bird, but will also reduce feed costs. The greatest heat loss is through the roof, with 35 per cent dissipating in this way, while draughts account for another 25 per cent loss. To meet these criteria, a house roof needs to be insulated to a value of $U = 0.5W/m/°C$ (MAFF/ADAS recommendation). There are a variety of insulating materials available, including compressed, fireproof wallboard and sheets of rigid, extruded polystyrene foam.

Bear in mind that the birds themselves will provide the warmth, while the insulation prevents its loss. Make sure that there is a maximum/minimum thermometer in the house to keep a check on the temperature range.

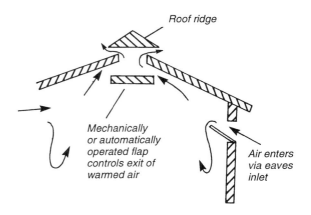

Figure 5.1   Ventilation in large houses may need mechanically or automatically operated flaps.

Ventilation

No bird will remain healthy for long if it is housed in a stuffy, badly ventilated house. There needs to be a continuous throughput of fresh air, balanced against the equally important need to conserve warmth. Getting the balance right is not always easy.

Large houses are best ventilated by a combination of eaves and ridge ventilation, referred to in the last chapter. Air comes in through the side vents or windows and, as it warms, rises to escape through the ridge outlets above. The number of windows depends on the flock density. Mechanically operated flaps can be inserted under the ridge outlet and these, together with the opening and closing of the eaves, allow a considerable degree of control. Fans may also be needed to drive the air though, but where flock density does not exceed seven birds per square metre, natural ventilation should be sufficient.

A house with natural light and ventilation. The wide pop-hole opens onto straw bales that provide a step as well as a feet-cleaning service. *(Onduline)*

Where an automated system is used, producers of Freedom Food eggs are required to have an alarm system in case the automatic ventilation should encounter problems. A back-up system generated by a 24 volt motor will provide a failsafe in the event of a power failure. Maintaining a pleasant environment inside a large house can be difficult, with a rapid build-up of ammonia presenting a problem. Dust and feathers may also contribute to lung infections in bird and man.

Layout

A convenient layout for a large house is to have a central egg-collecting passage with nest boxes on either side running the length of the house. The pop-holes are then along the outer sides. The distance between the pop-hole and the nesting area, in a direct line, should not be too great; it is important that the hen can find her way in and out easily. A confusion in this respect can lead to floor-laid eggs.

To obviate any problems of too great a flock density in any one area of the house, side panels can be incorporated to break up the length of the building, bearing in mind the need for pop-holes in each section. A belt system of feed delivery can still be catered for by allowing gaps in the side sections.

It should be emphasised that there are many different designs of house, and the layout described above is by no means the only possibility. Housing manufacturers have their own suggested designs or will build to the customer's own specification.

Lighting

Lighting is a major element in a large house, with a sufficiently sophisticated control system to allow for fine adjustments of day length and light intensity. A dimmer switch incorporated into the control panel is an effective means of controlling potential problems such as feather pecking in conditions that are too bright. It is also important not to have light shining into nest

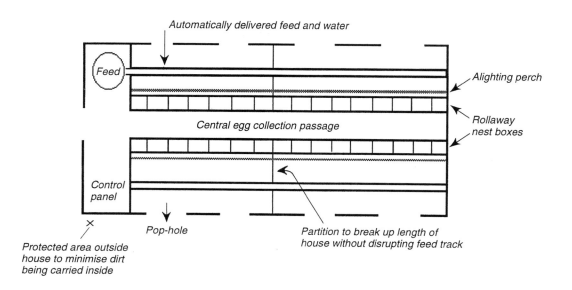

Figure 5.2   A convenient layout for a large static house.

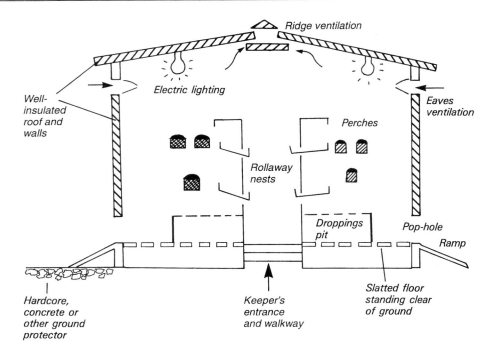

Figure 5.3   End-on view of a large perchery house.

boxes, which might encourage egg eating.

Forty watt tungsten bulbs or 6–8 watt fluorescent bulbs or tubes are satisfactory, as long as they provide diffused light. Clear bulbs throw harsh shadows which not only disturb the chickens but also contribute to floor laying. Light sources need to be placed 10 m apart. One light source is sufficient for up to 100 birds. Compact, low-energy, fluorescent lamps are now available, and are much cheaper to run.

Lighting is also required for the central egg-collection passage. This should be arranged in such a way that it does not shine into the nest boxes. It should also have its own circuit so that it can be switched on and off, as required.

Control panel

Electricity and lighting controls should all be in one area, neatly arranged for ease of access, and out of the way of the birds. A convenient area is near the central egg-collection passage of the house, or perhaps in the egg storage area. The salient points to bear in mind are as follows:

- Have a wooden panel placed securely and at a convenient height. Screw on all the appropriate switches and control systems such as the adjustable lighting control and ventilation controls, if used.

- Ensure that cables are protected against wet weather and condensation. PVC cables can also become brittle in the sun. Where there is a potential problem, use heavy-duty outdoor cabling, or place PVC cables inside black polythene water piping. Waterproof switches are also recommended.

- Ensure that all appliances which require earthing are properly earthed.

- Ensure that the correct fuses are used in each appliance or plug.
- Have all electrical installations checked by a qualified electrician.

Perches

Perches are required, arranged in such a way that easy access to nest boxes is provided. They can be placed parallel or stepped. With stepped perches, the height above the floor will vary, but the bottom one should not be more than 60 cm above the nearest surface. Reference was made in the previous chapter to recent research by Scottish Agricultural Colleges which indicates that the horizontal distance between perches, as well as the vertical distance from ground to perch, should not exceed 1 metre; otherwise the chicken may not be able to jump the gap.

There are many options for perches, including multi-tiered arrangements. The choice will be a personal one, depending on the type of house and the specific requirements of the packers. It is also a good idea to ask the advice of the regional egg marketing inspector when it comes to the positioning of perches and nest boxes to ensure that they meet the appropriate legal requirements, for free-range, organic or Freedom Food production.

Free-standing nest boxes with perch access. The roof is sloped to prevent perching there. *(Patchett Poultry Equipment)*

Nest boxes

The cleanest eggs are undoubtedly produced in houses which utilise 'rollaway' nest boxes. There is no great mystique about such a construction: it is merely a series of nest boxes with floors which slope downwards, away from the hen's entrance. The nest boxes are lined with nesting material, often with a strip of the material at the collecting side that acts as a cushion. After the egg is laid it rolls backwards to the collection point, landing against the protective cushion. The collection area projects further out than the back of the nest box so that the hen cannot reach the egg.

This helps to reduce the incidence of egg eating, a troublesome condition which can arise in even the best-run units. Netlon matting is often used in rollaway nest boxes.

The positioning of the nest boxes will depend on the design of house, although they should not be on the floor.[1] One producer found that nearly 25 per cent of his birds were not using boxes placed on the floor. When he was advised to raise the nest boxes by mounting them on the walls, the situation immediately improved. The birds were able to see the nest boxes, and floor-laid eggs fell to 2 per cent. Where nest boxes are raised, there must, of course, be alighting

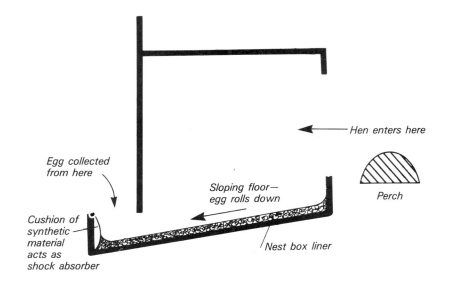

Figure 5.4 Details of a rollaway nest box.

Eggs that have rolled away from the nest boxes ready to be collected in the central passageway of a large perchery house. The nest boxes are lined with plastic netlon.

rails for the birds to land on. There should also be enough boxes for the flock, at least one for every five birds.

Feeders and drinkers

In a large house, the choice of feeder is sometimes an automatic flat chain feeder programmed to run several times a day. This is essentially a narrow trough with a chain powered by an electric motor. It is important that it is within easy reach of the birds, without being so high as to cause an obstruction between nest boxes and pop-holes. All feeders should be at the level of a bird's back so that wastage by the birds' habit of flicking out the food is avoided. The level of the food within the feeder should be kept fairly low for the same reason.

Alternatively, suspended feeders may be used. These may be manually filled or automatically replenished by an overhead auger from a bulk feed bin. This is powered by a motor and transfers feed from the bin, via delivery tubes, to the feed pans. With a bulk bin, there will be the advantage of reduced feed costs, as it is more expensive to

Here, a protected scratching area verandah acts like a bridge over a drop in the ground. The silo on the right provides feed for the automatic feed system in the house.

buy feed by the bag than in bulk. Suspended feeders are available in galvanised metal or plastic, while the feed pans of an automatic system are made of high-quality plastic.

An automatic watering system is appropriate for a large house. This will consist of a header tank fed by the mains supply, along with feeder pipes and drinkers. The drinkers can either be a series of bell drinkers or a system of nipple drinkers. The most common are suspended bell drinkers with a valve to maintain a constant depth of water at a pressure of 7psi. The valve also enables the water to be switched off when the drinker is removed.

It is important that feeders, drinkers and perches are above a slatted droppings area because this is where most of the droppings and mess will be produced.

Needless to say, the systems should be checked regularly to ensure that they are clean and delivering properly.

Freedom Food housing requirements

In addition to the basic free-range requirements, many large and small producers are now also meeting the requirements of the RSPCA's Freedom Food standards. It is necessary to apply for registration and if, following inspection, this is accepted, the description Freedom Food can be used as a marketing description. Housing is required to comply with the following:

* Stocking rates should not exceed seven laying adult hens per square metre of floor space. In a house with a perching area over a droppings pit, the maximum density can be up to 11.7 hens per square metre of available floor space in order to allow for the natural clustering of birds. If a multi-tier system is used, where overhead perches/platforms

with feeders and drinkers provide enough space for at least 55 per cent of the birds to perch, then a stocking rate of up to 15.5 hens per square metre of available floor space is allowed. Most producers seem to abide by the first option of seven (or even fewer) adult hens per square metre.

* The house should have a notice at or near the entrance indicating the salient points of the system, and of the stocking rate in relation to available floor space. The following is a fictitious example:

FOXFREE FREE-RANGE
Sunny Farm
Chickfold
House Size: 70 × 10.5 m = 735 m²
Number of Birds: 5,000 = 6.80 birds/m²
Date housed: 6/4/1996
Supplier: Happy Hens
Breed: Browny Egger

* Housing and equipment must be designed so that all the hens can be clearly seen.

* There must be enough pop-holes to ensure that all the hens have ready access to the outside. Each must be at least 45 cm high and 2 metres wide to allow several birds to exit at the same time. There should be one for every 600 birds, and there must be no obstructions that would prevent birds seeing the pop-holes.

* Provision must be made to provide an adequate house temperature, taking into consideration the upper and lower critical temperatures of the birds, given their size, feed intake, group size and climatic factors. In winter conditions, a shelter providing sufficient protection from wind, rain and snow and of sufficient size to accommodate all the hens must be available. In summer conditions, a shaded area must be provided to avoid heat stress. Overhead cover is also required to reduce fear reactions to overhead predators.

* In houses, poultry must have access, at all times, to a well-maintained litter or well-drained resting area. Litter must make up at least 33 per cent of the floor area and be maintained in a friable condition. The minimum depth is approximately 10 cm. If the dustbathing environment is provided outdoors, access to this area must be allowed for at least four hours a day.

* All floors must be designed, fitted and maintained to avoid distress or injury to the hens.

* Toxic substances must not be used on any surfaces accessible to the hens.

* Perches should be approximately 4 cm wide, be non-slip and have no sharp edges. There should be a gap of no less than 1.5 cm on either side of any perch so that there is no risk of birds trapping their claws. A minimum of 15 cm of perch space per hen should be allowed, including the alighting rail in front of a nest box. Where perches are incorporated within or attached to a slatted or mesh-floored area, there must be some perches above ground level to allow hens to escape from aggressors. Perches must be positioned to minimise fouling of any hens below and should be over a droppings pit.

* One nest box for every five hens should be provided, or 1 square metre of nest box area per 120 hens. These should be provided with a suitable nesting material which encourages nesting behaviour.

* All mains electrical installations must be properly earthed and inaccessible to the birds.

* Ventilation rates and house conditions must provide sufficient fresh air at all times. Aerial contaminants should not reach a level where they are noticeably unpleasant to a human observer. Specifically, inhalable dust must not exceed 10 mg per volumetric metre, and

ammonia should not exceed an average of 25 ppm over any eight hour period.

Using outbuildings

Sometimes existing farm buildings can be converted for free-range production. Old mushroom houses, pigsties and other buildings have all been used to good effect. One enterprising farmer bought a number of old caravans which were no longer roadworthy. After stripping them out and providing nest boxes and perches, he used them as moveable houses for a number of small free-range flocks. Moving them around was comparatively easy with the aid of a tractor, as they were already equipped with towbars.

The key factors in the utilisation of existing structures is that there should be an adequate level of insulation and ventilation. Old buildings are notoriously draughty. Floors, doors and windows should all be checked to make sure that there are no access points for rats. The base of doors is a particularly common entrance point. If this area is worn, it is a good idea to reinforce it with a metal panel, as shown in Figure 5.5.

Buildings can also be refurbished with insulation boards, which are widely available from DIY suppliers. Steeply pitched roofs on old farm buildings may need to be fitted with false ceilings if the heat loss through the roof is high. At the same time, if a false ceiling is inserted, it is important that it is not too low; otherwise there could be ventilation problems. Existing air outlets should also be left unrestricted.

If the floor is concrete, ensure that it is painted with a sealant. Litter such as sawdust, wood shavings or treated chopped straw will be needed. Bear in mind the remarks made above about the dangers of dust and lung infection. Floor litter is also an incentive to floor-laid eggs so a careful watch will need to be made for this.

A droppings pit should be provided under

A farm building converted for free-range birds. A slatted floor has been inserted, allowing space for a droppings pit below, hence the high pop-hole. The ramp is wide for ease of exit and the area around the house has wood chips to keep the hens' feet clean before they go in.

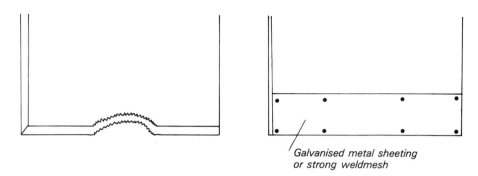

Galvanised metal sheeting
or strong weldmesh

Figure 5.5   Reinforcing worn areas against vermin entrance.

all the perches, while feeders and drinkers should be placed on a slatted area. Fresh litter will need to be added as required, to keep it in a friable condition. The litter is cleared away at the end of flock lay. It can be composted until well rotted. If subsequently bagged, it can be sold as a soil conditioner.

Where a refurbished or other static farm building is the housing, a range of free-standing nest boxes with alighting perches may be appropriate. It is also possible to buy sections of rollaway nest boxes that can be erected to the best advantage, depending on the type of building. Suspended tube feeders connected to an automatic system of delivery from a feed auger can be set up, as well as an automatic water system with header tank, pipes and drinkers. As referred to above, these can be suspended bell drinkers or nipple outlets.

Mains electricity will be needed. Supplying it to outbuildings is normally not difficult, and indeed it may already be there.

Polytunnel housing

A form of low-cost housing which is popular with some producers is the polythene tunnel house. The price compares favourably with other forms of housing and it is moveable. It

can, for example, be erected straight onto the ground, without the need for a concrete base. However, the question of how to deal with rats must inevitably arise with this situation.

A typical structure has a galvanised steel frame attached to concrete anchor points, and bolted together with locking nuts and bolts. Timber side frames and wall sections enable pop-holes to be incorporated, while the structural hoops can be spaced at 2 metres distance to allow the pop-holes to meet Freedom Food standards. End frames allow the positioning of doors at either end, and guttering can also be attached.

There are normally two layers of polythene with insulation in the middle. The inner skin is white, allowing reflection of light, while the outer layer is coloured plastic of a heavier gauge. This is treated against ultra-violet light degradation, and will normally need to be replaced every five years.

Perches and nest boxes can be provided in a polytunnel house, as well as lighting, from a 12 volt battery or mains system. If an automatic delivery feed system is used, mains electricity is esssential. This can be provided via a cable which runs through a time switch and dimmer which itself plugs into the mains.

The West of Scotland Agricultural College carried out a research project to assess the viability of polythene free-range housing.

The house was composed of a double layer of white polythene, stretched over a galvanised steel framework, with an insulating air pocket between the two layers. It was placed directly on an earth floor so that no site preparation was necessary, and it was easily moveable to a new site when required. Six hundred 17-week-old ISA Browns were introduced to the house, with the range area being divided into two so that the birds could be released alternately on to one side and then the other. The stocking rate at any one time was 770 birds per hectare, considerably below the EU regulations requirement of 1,000 birds per hectare. The birds were fed a 17 per cent layers' mash from a mechanical chain feeder five times a day, resulting in a mean figure over the year of 137 g per bird per day. Eggs were collected three times a day. A summary of the results from 20 to 72 weeks of age is given in Table 5.1.

Table 5.1   Egg production in polythene housing

Hen-housed average	296 eggs
Floor eggs	2.68%
Mean egg weight (from 33–62 weeks)	62.5 g
Percentage eggs graded 1, 2 and 3	40.8%

Source: Linda Keeling and Peter Dunn, Polythene Housing for Free-Range Layers: Bird Performance and Behaviour, Scottish Agricultural Colleges. Note No. 41, February 1988.

Egg production was good, with a hen-housed average of 296 eggs per bird, but egg sizes were low. This was thought to be due to the birds coming into lay early because of the considerable light entering the house. Coloured polythene, as described earlier, would have prevented this. It is important not to enourage laying until the body frame and eating capacity of the birds are sufficiently developed for egg production.

Housing the organic flock

Smaller producers are in a good position to provide organic eggs, for they are not dealing with the flock sizes and densities of the large enterprises. Some of the larger moveable houses referred to in the previous chapter are ideal for this, as well as for the rearing of table birds in small numbers.

Houses used by those who are part of the Soil Association Organic Symbol Scheme, for example, must meet the following requirements:

- A maximum of 500 birds (layers or table birds) in a house. With more than a normal social group of 100 birds, the number and distribution of feeders, drinkers and other facilities must be adequate to allow development of social groups within the unit.
- A maximum stocking rate, in fixed or mobile housing, of six birds per square metre (layers). For table birds, ten birds per square metre in fixed houses and 16 in mobile houses. Minimum perch space is 18cm per bird, with one nest box for every six birds.
- Pop-holes: a minimum of 4m length per 100 square metres of floor area.
- There must be additional shelter against wind and rain and outside drinkers are recommended.
- A maximum slatted floor area (per cent of floor area) of 50 per cent.
- Stocking rates should not exceed 1,000 laying birds per hectare, or 2,500 table birds per hectare.
- In the design of poultry enterprises, preference should be given to mobile houses, as these allow for greater flexibility of management and the ability to integrate the poultry operation into the organic farming system.

Reference was made earlier to the availability of a new, lightweight range of mobile housing for free-range and organic flocks. (See the photograph on page 115.)

This system is the ideal, in my view.

References

1. Dalgety Poultry Review.

CHAPTER 6 **Land Management**

The poultry-man who adopts free-range methods should move the houses frequently and keep the grass short.

Leonard Robinson, 1948

The key to successful free-ranging is good land management. This applies to the nature of the land itself, the degree of shelter it offers, how it is fenced to deter predators and how pasture is managed and maintained. Traditional free-rangers have always known this, but some of the first large-scale commercial enterprises in the 1980s failed to appreciate it, thinking that the house was the only thing that mattered. They ended up with houses so large and badly designed that some birds never found their way out. Those that did stayed in the immediate vicinity of the building because no provision had been made for shelter in the pasture, and the area immediately around the houses became devoid of vegetation. Complaints from welfare organisations and consumers that this was not real free-ranging soon had its effect, and new standards were introduced to improve conditions. There is still some way to go, however, in convincing some producers that the grazing area matters.

The nature of the land

It is no coincidence that the great free-range egg-producing areas of the past were in relatively mild areas blessed with free-draining soils. The Lancashire sands and Wiltshire chalks were ideal, providing land free from boggy areas, although in very hot summers there could sometimes be problems of grass scorching in particularly thin-soiled areas. The ideal is to have adequate drainage to prevent waterlogging, while ensuring that the bulk content and fertility of the soil are sufficient to retain and provide enough water and nutrients for a healthy growth of grass. The balance is not always easy to achieve, but it is better to start with a well-drained soil than a heavy one.

It is possible to improve drainage, but this can be an expensive option. Where the water table is naturally high, the only long-term solution may be to plough the whole area and excavate drainage ditches, as has been done in parts of the Fens. Such an exercise would not be cost-effective.

It is possible to install permanent drain pipes – and this should certainly be the case in the area around a permanent house – but it is expensive for the rest of the land, and may also be difficult to justify in terms of the likely returns. In waterlogged areas, where the water table is not normally high, the problem may be one of 'panning'. This occurs when the top few inches of a soil have become so compacted that the surface pan holds the water, without allowing it to drain through. It is usually clay soils which are affected in this way, although lighter soils may become panned if overuse has led to excessive compaction. The solution here is to mole plough the area. The 'mole' in question is a tractor attachment which breaks up the hard surface and forms a series of tunnels in the subsoil. There are agricultural contractors who will undertake this.

Once the area is ploughed, an application of lime in the autumn will help to flocculate the soil, the process by which tiny particles of clay clump together to make bigger particles, thus providing larger air spaces for more efficient drainage. The addition of nitrogen from a source such as calcified seaweed later in the season will increase the fertility while not creating a problem of leaching into the water courses. A new grass ley mixture can then be sown after the land has been harrowed and levelled. Once the grass has grown and established, it is ready to provide pasture for the birds.

Where waterlogging is only apparent in small areas, the easiest solution is to dig a hole and place clinker at the bottom to make a soakaway. The point has been made elsewhere in this book that, where fixed houses are used, there should be a protected verandah to provide extra shelter, as well as a shingle or slatted area by the pop-holes to prevent muddy conditions.

Some of the large free-range producers today are still not giving pasture rotation the importance it deserves. A common fallacy is the belief that a maximum of 1,000 birds per hectare is too light to cause damage to the turf, or to bring about a build-up of disease-causing organisms.

The fact that commercial point of lay birds are innoculated against a wide range of diseases, and are kept for a shorter period of time than they were traditionally, does give them a degree of protection that flocks did not have in the past, but producers are wrong to rely entirely on these factors. Disease organisms can and do mutate, and the traditional practice of land rotation in order to break the life cycle of pests and thus bring about their demise is well proven.

The physical structure of the pasture and topsoil can also be damaged by overuse. A problem that I recently detected in some large units is that of 'mole-hilling'.[1] Here the birds scratch away the turf in order to make

Sheep help to keep the grass cropped, making fresh, short growth available for the hens. The hens have scratched the turf to make dust baths, eventually producing 'mole-hill' undulations.

dust baths. This is entirely natural behaviour on the part of the hens, but where they are kept in large numbers on the same area of land, the density of dust baths can become so great that the pasture develops undulations. This, in turn, damages the shorter growing grasses needed by the birds, while allowing coarser grasses to gain a foothold. Once pasture has degenerated to this condition, it

One of the author's Lohmann Browns taking a dustbath. This is an instinctive pattern of behaviour which helps to keep the plumage free of parasites.

should be ploughed, rolled and harrowed ready for a new sowing, while the birds enjoy new ground.

Flock density regulations

The EU free-range regulations require only that the land be 'mainly covered with vegetation', without specifying the type or condition of plants. The Freedom Food directive is more demanding and specific:

- A grass sward must be maintained over the grazing area, with active management of damaged ground.
- The stocking rate must not exceed 1,000 birds per hectare available to the hens over the flock life.
- Land used for arable cropping shall not be regarded as acceptable vegetation and shall be excluded from calculations for stocking density.
- Where there is a build-up of parasites or disease on free-range land, rotational grazing or other disease control measures must be applied.
- If rotational grazing is used, a minimum of one-sixth of the total range area must be available at any one time.

This is all right as far as it goes, but there is nothing which states that ground *must* be used in rotation. The use of the phrase 'or other disease control measures' is too vague and open to abuse, such as a permanent reliance on chemical methods of control.

The organic standards of the Soil Association are as follows:

- The land to which the birds have access must be adequately covered with properly managed and suitable vegetation. Recommended are grass/clover leys based on fescues and other grasses which tend to tillering rather than leaf length. Conditions that favour the development of natural dusting areas are advised, with companion grazing with sheep for sward management.

- Stocking rates should not exceed 1,000 birds per hectare.
- Pasture must be rested to allow vegetation to grow back. In the case of layers, the fallow period should be at least 9 months. With table birds, this should be at least two months per year, and in addition for one year in every three years. These requirements do not apply where under 50 birds are free to roam rather than being in pens.

Sheltered areas

Land which is to support free-ranging chickens adequately must be sheltered. This does not refer to the obvious need for a house, but to factors such as the availability of trees, walls, windbreaks, shelters and hedges. The chicken originated in subtropical forests where trees provided cover and protection from wind, rain, sun and predators. It is out of its natural environment in a wide open field, and tends to stay near the house. Consequently, the grass immediately around the house soon deteriorates, while that further afield is barely touched. Although EU regulations demand that the stocking density does not exceed 1,000 birds to the hectare, the actual density in the limited area the birds tend to frequent is usually far higher than this, unless they are given positive encouragement to range further afield.

The Scottish Agricultural Colleges have conducted a survey to investigate the distribution of land usage and the number of birds going out onto range. By plotting the location of birds at one-hourly intervals, they discovered that 55 per cent stayed in the area around the house, thereby using 8 per cent of the total area available. The mean distance of birds from the house was 8.3 m out of a possible 80 m.[2] See Figure 6.1.

Weather is obviously a factor which influences chickens in using the range, with warm, dry conditions being more conducive to ranging than wet, windy ones. Hot sun is not particularly liked, but warm, relatively shaded conditions are ideal, reminding them of their jungle origins. However, the shelter/security factor is not one that should be neglected. My own observations are that chickens will range much further afield if there are plenty of trees on the site. Keeping chickens in an orchard proved to be the ideal, with a far more uniform distribution over the whole site than when I had them in an open field.[3] If a site is properly fenced against foxes, the provision of trees is an excellent inducement to the chickens to range over the whole protected site. Trees also provide much-needed shelter from wind, although any which are too near the perimeter fence may need to be cut back if they are likely to provide a launching pad for chickens to flutter over.

Trees are not the only form of shelter, of course. Wattle hurdles, hedges, banks, straw bales and stone walls all provide protection from the wind, as well as the feeling of security which is so necessary for chickens if they are to forage all over the available ground. It is interesting to note that a free-range unit which was set up in an old three-acre walled garden experienced no problems in getting the chickens to range all over the site. There were several trees, and the high stone walls gave an overall feeling of enclosure.[4]

Moveable shelters with outside drinkers are recommended for use in hot weather. Such shelters need not be elaborate or expensive constructions, merely something that will provide shade from the hot sun and help to keep the water supply relatively cool. Registration with Freedom Food requires an overhead cover shelter for winter and summer conditions, as well as to reduce fear reactions from overhead predators.

On a small scale, a simple temporary shade shelter is easily made by banging some posts into the ground, placing wire netting on the top and covering it with black plastic weighted down with turf or a few leafy branches.

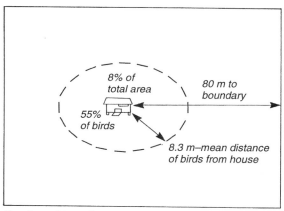

Drawing not to scale
Based on information from Scottish Agricultural College

Figure 6.1   Mean distance of movement.

Fencing

Traditional wisdom has always valued the adage, 'Good fences make good neighbours.' As far as the free-range enterprise is concerned, good fences also make good sense! Without them, the depredations of the fox would make a mockery of any effort to keep poultry. In addition, wandering dogs, feral cats, mink, badgers, even polecats in some areas, may prove hazardous to chickens.

Fencing is also required to control the birds' access to pasture, so it is appropriate to regard fencing in two ways: external or perimeter fencing to keep out predators, and internal or pasture control fencing to restrict the movement of chickens to certain areas.

Perimeter fencing

This should separate the poultry farm from the outside world. There is no easy or cheap way of doing this and it is one of the most substantial costs facing anyone thinking of starting a free-range poultry farm.

A determined fox can scale a fence, although to go back over it carrying his prey may prove too much for him. However, that is a hollow victory for the poultry-keeper if the fox has already killed several birds. The vixen, hunting for her young ones in early spring, will often kill many birds, biting off their heads, then making off with one, leaving the other corpses behind. A 2 m high perimeter wire mesh fence of will provide adequate protection, as long as there is a further overhang of 30 cm placed at an angle of 45° to the vertical. The overhang should project outwards to repel boarders. Some poultry keepers claim that to have the top strained and projecting outwards is unnecessary, and that merely having a 30 cm section, loose and unstrained at the top, will dissuade a predator because it will not provide sufficient support for scrambling up it.

Wire mesh netting with 50 mm holes in the mesh will deter foxes. Mink can still climb a fence and wriggle through holes of this size. However, they are more likely to be a problem in areas close to rivers. In such situations, trapping may need to be considered, and the Ministry of Agriculture

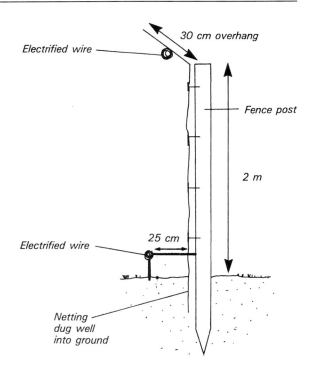

Figure 6.3　Electrified perimeter fencing.

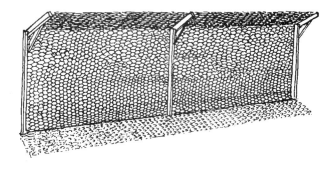

Figure 6.2　A 2 m fox-proof, non-electric perimeter fence, dug into the ground and with an overhang to prevent climbing.

should be consulted for advice on suitable traps and procedures. In some areas it is possible to borrow traps from the authorities.

The bottom of the wire mesh needs to be dug well into the ground to prevent predators from pushing their way underneath. The provision of an electrified wire will give added security to the fence, in this respect, if it is placed 25 cm above the ground and 25 cm out from the mesh. Placing a second wire near the top, or just above the fence, 10 cm out from the mesh, will make the barrier virtually invulnerable. The details are shown in Figure 6.3.

A high-power mains fencer unit is recommended for a permanent fence of this kind, because it needs less maintenance than a battery-operated one. Wiring should never be connected directly to the mains supply

An existing post and rail fence covered with galvanised poultry netting, and electrified by two strands of wire at the top and bottom.

horizontal lines of heavy gauge polythene/stainless steel electroplastic twine, with non-conducting polythene verticals and bottom horizontal strand. The gauge of the fencing is important, with smaller holes at the bottom, increasing in size to the top. This ensures that the chickens, particularly young pullets, do not become entangled if they should happen to touch the netting. The fence is tensioned with straining post guys and pegs, and powered from a mains- or battery-operated unit. The latter is normally a 12 volt rechargeable car battery.

No system is without its problems, and the chief ones here are shorting and failure of the power supply. Shorting can occur if the grass gets too long where the fence is positioned, so it is important to keep that area mown. A battery-powered unit should be regularly checked in case it needs recharging. A neon tester or electric voltmeter for checking the state of electrification is highly recommended.

When the fence is moved, the posts are pulled up, the net is rolled up and the fence re-erected on the new site. A problem which may arise at this time is tangling of the netting. It is much easier for two people to move the fence and to ensure that it is evenly rolled. To avoid tangling, it is possible to use

and the advice of a specialist should be sought before installing such a system. There are many suppliers of electric fencing who are experienced in the needs of poultry keepers, and who supply fences which conform to the British Standards Safety Requirements.

If an existing fence is electrified, the cost is about half that per metre of installing a new one.

Pasture control fencing

This is to control the access of birds to certain areas of pasture, rather than to keep out predators. It does not need to be particularly high – 90 cm is usually sufficient – but it should be easily moveable and re-erected as required. Electrified netting is ideal for this purpose; it is also suitable for incorporation in a permanent perimeter fence. It is essentially a series of lightweight plastic fencing posts with metal spikes which are tapped into the ground. These are non-conducting and are purely for support, with one being placed every 3 m or so. The netting is normally made up of eight

Electric poultry netting. *(Renco)*

a system of netting that is coiled around an applicator that is then wheeled along to the appropriate site.

Some large producers also use a strand of electric wire inside the house if there is a problem with floor-laid eggs. This is an abhorrent practice that puts the convenience of the producer before the welfare of the birds. It is one thing to have an electric fence outside in the field, for the protection of the chickens; it is quite another to put it in the place that they normally associate with security. It can cause extreme stress. A good management system of regular inspection and attention to lighting and floor and litter conditions will produce a minimum of floor-laid eggs.

Paddock rotation

A field is often regarded as a permanent and unchanging entity. In one sense this is correct, but if the grass is regarded as a crop, it is obvious that it is temporary and must be managed properly on a seasonal basis. As referred to earlier, chickens cannot be allowed to range on the same piece of land indefinitely; otherwise the grass will deteriorate, there will be a gradual increase in the incidence of pests and parasites, and the flock will succumb to health problems.

ADAS recommends that chickens should generally be moved to a new area of vegetation every 4 to 6 weeks, if a system of paddock grazing is used. With a moveable house this poses no great problem; it is simply a matter of transporting it by hand or tractor, depending on its size, and re-erecting the pasture control fence of electric netting around the new area. Where only a small number of birds are kept in a house with combined run, it may be more appropriate to move it every day or every few days. If the flock is ranging in numbers well below the official limit and has a considerable expanse of pasture at its disposal, it may not be necessary to move it as frequently. For

example, a house may have a field on either side of it, with these two areas being used alternately. It is a question of relative scale, stock density and common sense, but with a commercial unit, the ADAS recommendation is a good general guide.

Where fixed houses are used, the birds' access to new grazing is more difficult to arrange, but a portable electric netting fence is an effective means of controlling the flock's ranging once it is outside.

Paddock management

After a flock has been moved to a new area, the old pasture should be raked if there are any patches of compacted droppings, and then 'topped' to cut down taller grasses which may be producing seed heads. An alternative is to graze sheep, cattle or goats on the site; they will do all the necessary topping.

Some free-range units rent out such land to farming neighbours who require extra grazing, while others graze sheep at the same time as the birds.

On a large scale, topping can be done with the traditional farm equipment of tractor and cutting bar. A ride-on garden tractor with the blades set at maximum height is also effective. For small areas, an ordinary hand lawn mower with the blades set high will do the job.

Mention has already been made of the need to keep grass short near electric fencing. Regular mowing is the only way to ensure this and a ride-on or hand garden mower is effective in keeping a strip close to the fence clear of tall growth. Grass strimmers have also been used to good effect. Remember to switch off the electric fence before mowing around it!

It may also be necessary to mow the area of pasture currently occupied by the birds, if it is growing more quickly than their foraging action can keep pace with. Tall grasses are not eaten by the birds and rank growth,

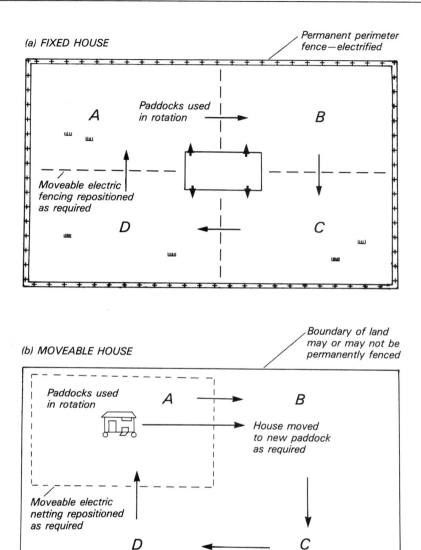

Figure 6.4 The principles of free-range.

particularly if wet with rain or dew, is a positive disincentive to them. Those which do brave such conditions get their legs and bottom feathers wet, making eggs which are subsequently laid wet and dirty.

Once birds have vacated a site, and if other livestock is not being grazed on it, it is a good idea to dust with lime, particularly if the ground tends to be on the heavy or acidic side. There is some dispute among poultry experts whether this practice helps to deter disease-causing organisms and clear up parasitic infection of the ground. Traditional wisdom has always encouraged it – and what is certainly true is that heavy clay soils are less likely to compact and puddle if treated with lime. A reduction in the number of waterlogged areas will also reduce the incidence of snails which act as intermediate hosts to parasitic flukes and the coccidiosis causing organisms. Liming is a practice I always follow to good effect with my own pasture rotation management.

If an area of pasture has deteriorated badly, it may be appropriate to plough it up, harrow it and then reseed it with a new ley mixture. Once it is ploughed, test the soil to determine its pH level of relative acidity or alkalinity and apply lime if necessary. Leave it to weather for a time, then harrow it to break down the clods of earth ready for seeding. All these activities can be carried out by a specialist contractor, if you are not in a position to do it yourself.

Special ley mixtures for free-ranging poultry are available from some specialist seed suppliers. These mixtures are made up of shorter perennial grasses which are more suitable for poultry than the longer grasses which are usually found in other leys. As a general rule, 50 gm per square metre, or 500 kg per hectare, of seed will be required.

A new ley pasture will not generally require feeding in its first year. In the second year, it can be given some nitrogen fertiliser such as environmentally acceptable calcified seaweed.

References

1. Author's observations at three large units, 1996.
2. Linda Keeling and Peter Dunn, *Polythene Housing for Free-Range Layers: Bird Performance and Behaviour*, Scottish Agricultural Colleges Research and Development Note No. 41, February 1988.
3. Author's unit at Widdington, Essex, 1975–88.
4. *Poultry World*, June 1987.

A free-range egg farm with several houses from which the chickens have access to the same pasture. Straw is used to protect the worn areas of turf around the houses. *(Onduline)*

CHAPTER 7 Feeding

If the last portion of food necessary to satisfy a bird's appetite is missing, it is egg production which is likely to suffer.

Keith Wilson, 1949

The chicken has no teeth, and food is swallowed whole after being picked up by the pecking action of the beak. It is able to tear off pieces of food such as the tips of grass, but there is no biting or chewing action. Once swallowed, the food passes down the gullet into the crop, which is like a temporary storage pouch. If you pick up a chicken just after it has eaten wheat, you can feel the individual grains in the crop, which is in the breast area.

In the crop, food is moistened and softened by secretions from the mouth, gullet and crop itself, making ready for subsequent digestion. From here, as soon as there is room, food moves down into the proventriculus, where gastric juices containing hydrochloric acid and the enzyme pepsin begin to break it down. It then passes into the gizzard.

The gizzard is a bag equipped with strong muscular walls that contract and relax. It

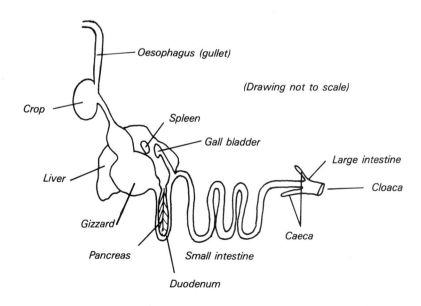

Figure 7.1 Digestive system of the chicken.

contains small particles of grit, or tiny stones that the bird has swallowed. These act rather like mill stones to grind up the food, and it is here that the real work of breaking down particles such as grains takes place.

The food passes along to the small intestine, via the duodenum, where enzymes and secretions from the liver, gall bladder and pancreas continue the process of digestion. Bile passes from the liver, is temporarily stored in the gall bladder, then moves in to emulsify fats in the food matter, enabling enzymes to complete their digestion.

The food, broken down into its constituents, is then absorbed into blood capillaries called villi that line the wall of the small intestine. From here, blood transports it to every part of the body where, in conjunction with oxygen carried from the lungs, it provides nutrients and energy for the whole system.

Any unabsorbed particles pass from the small intestine into the large intestine. Urine filtered by the kidneys is also delivered into the latter part of the large intestine via ureter tubes. Projections called caeca reabsorb some of the water, but the waste urine and faeces are both expelled from the cloaca. The whitish part of the droppings is the urine.

Feed requirements

The metabolism of the hen – like that of any other creature – includes a wide variety of activities. She grows, repairs her tissues, grows healthy new feathers, breathes, moves about, lays eggs, clucks, pecks and scratches. The list is almost endless, and these activities would soon be curtailed if food was not available to power them.

The basic requirements, apart from water, are proteins, carbohydrates, minerals and vitamins. All these foods cater for the overall metabolism of the bird but, as a generalisation, proteins may be said to cater

primarily for growth, carbohydrates and fats for energy, and minerals and vitamins for health. This is an oversimplification, but as a working definition for preparing a hen's feed ration, it is an adequate one.

Chickens have a small digestive system and metabolic requirements cannot be met by feeding large quantities of feeds which are low in proteins, energy, vitamins and minerals. The ration must be a balanced one, available at frequent intervals.

Proteins

Proteins are complex substances found in both animal and vegetable sources. Until recently meat, blood and bonemeal were utilised in poultry feeds, but following the BSE crisis in cattle, these are now banned. The main vegetable sources are soya and other beans, maize and other cereals, and sunflowers. Full fat soya is preferable to extracted soya because there are claims that the solvent used to extract soya oil from full fat soya can be carcinogenic.[1] There are also concerns that new, genetically altered maize from the USA could compromise the treatment of disease in animals and humans because it contains a gene that is resistant to Ampicillin antibiotic.[2]

Most free-range and organic rations are now based on plant proteins.

Proteins are made up of constituents called amino acids. There are about a dozen of these but the most important are lysine, methionine and tryptophan. A hen is capable of synthesising most amino acids from other food constituents, but these three must be taken in directly every day. The average laying bird will require a daily intake of 900 mg lysine, 430 mg methionine and 200 mg tryptophan. Linoleic acid is also essential in maintaining egg size, and the amount of this can be varied in order to produce eggs of a certain size.

The overall protein requirements are 18 to 19 g per bird, per day, although this may increase at certain times during the egg laying period. Most compound feeds declare the protein content on the sack, and this is normally between 16 and 18 per cent. A 16 per cent ration would require a higher daily intake than the 18 per cent ration. Commercially, compound feeds are often fed on an ad-lib basis, where birds can help themselves from feeders placed in the poultry house. Alternatively, an automatic system delivers feed several times a day. The compound feed intake needs to be balanced with a grain ration. This not only meets energy requirements but also keeps relative costs at an acceptable level. Compound feeds are far more expensive than cereals.

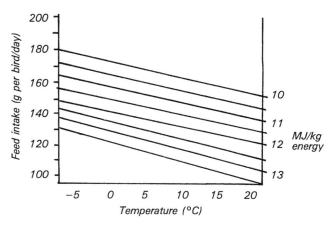

Source: Tony Warner, ADAS Reading, ATB Course, 1987.

Figure 7.2   Feed intake of free-range birds in relation to temperature.

Energy foods

Carbohydrates and fats are the important energy producers and these are found primarily in cereals and beans. Energy in feedstuffs is referred to as its metabolisable energy (ME), and this is measured as megajoules per kilogram (MJ/kg). A laying bird will normally require 11.5 MJ/kg, but temperature has an important bearing on this.

The optimum temperature for a layer is 21°C, a reminder of its warm climate origins. In winter a free-ranging bird is asked to produce eggs in temperatures which can go down to below freezing point. For every 1° fall in temperature from the optimum, a laying bird will need an extra 4.2 calories. This necessitates a considerable increase in energy-producing feedstuffs and the feeding of extra cereals as a scratch feed in winter. Ensuring that a house is well insulated will go a long way towards keeping the layers warm, but there is no alternative to extra winter feeding. Extra compound feed will also make up the deficit, but the cost of this compared with that of grain makes it financially unviable, bearing in mind that the cost of cereals is lower than that of compound feeds.

Figure 7.2 shows how increasing the

MJ/kg energy value in the diet by feeding higher levels of grain prevents the wastage of protein which results from higher intake of compound feeds.

Minerals

Adult chickens require the normal range of minerals that most living organisms use, but the crucial ones are calcium and phosphorus, which play an important part in bone structure and egg shell quality. Compound feeds contain both these minerals, with around 3.5 to 4 per cent calcium and about 0.3 per cent phosphorus. If such feeds are given it is not normally necessary to add extra limestone, as long as the calcium is readily available to the birds. If powdered mash is used, there may be a problem of availability, as the following indicates.

A 1965 study revealed that finely ground limestone sifted through the feed and settled at the bottom, so that birds did not always take in their required amount.[3] The result was thin and brittle shells, with a breakage

(Above) Layers' pellets being put in a plastic feeder which can be suspended if necessary.

A dustbin with a lid (left) provides excellent feed storage for the small flock.

level of 10 per cent, although the 36 week old birds were being given a 16 per cent protein feed with a 3.8 per cent calcium level. The problem was solved when coarse ground oystershell was introduced into the feeding system.

Salt is an important part of the diet but should not exceed a level of 0.4 per cent. Too little will adversely affect growth and egg production, while too much will cause excessive drinking, leading to digestive problems and loose droppings. Again, compound feeds contain the optimum level, normally 0.38 per cent.

Trace elements are small quantities of essential minerals and include zinc, iron, copper, selenium, iodine and molybdenum. They are normally included in compound feeds, where they are referred to as ash.

Vitamins

Vitamins are organic compounds, as distinct from the inorganic minerals. Many are available via the grass-grazing activities of the chickens, while vitamin D is produced through the action of sunlight. Small quantities of essential vitamin supplements are included in compound feeds. Vitamins can be added to the drinking water if there should be a sudden need. The effects of stress, for example, can be mitigated by giving the flock a multi-vitamin supplement.

Compound feeds

Compound feeds are rations which combine all the basic nutrients that are necessary in the bird's diet. Based on cereals, they contain mineral, vitamin and trace element supplements to provide a balanced formulation for dietary needs. Often referred to as 'mash', the feed is available in powder, crumb or pellet form. Most feed suppliers now supply compound feeds specifically for free-range flocks and these are given in association with a separate grain ration. As a general guide, each bird will consume 130 g of layers' meal a day, although this will obviously vary depending on breed and circumstances.

Commercial free-range units use compound feeds in the form of mash because this is the cheapest. It is suitable for use in automatic chain or pan feed systems, or manually filled tube feeders. The pelleted form is often the choice of small poultry keepers because, although more expensive, it is easier to handle and more convenient to feed. There are some who say that pellets produce over-fat birds, but where they are balanced by adequate grain and exercise, this is not the case. It is as well to remember the findings of Hull and Scott who showed that it was birds with a diet lacking calcium which tended to over-feed.[4]

Consumer reaction to what was perceived as an unacceptably high level of additives in poultry feeds, together with a general growth in the free-range market, has resulted in the production of compound feeds with more natural ingredients. Artificial yolk pigmenters such as canthaxathin, for example, are used in intensive systems because battery birds do not have access to grass, and their eggs would otherwise have pale yolks. Free-range egg yolks are normally well coloured, but in winter when the grass is not growing, they may be pale. Feeds often contain a certain amount of grassmeal, lucerne or maize products for this.

In the late 1970s, I wrote to virtually every feed company in Britain because I was preparing an article on poultry feeds for free-range poultry keepers. My letter asked, simply, whether they would let me know what ingredients were used in their feeds. With the exception of one company, they refused. When I asked the Ministry of Agriculture for their help, they told me that there was no requirement for feed ingredients to be disclosed. If only there had been! A lot of subsequent problems, such as the outbreak of BSE in cattle, might have been avoided. (It is thought that the practice of using dead, diseased animals as sources of proteins for livestock feeds might have been the cause of BSE in cattle.) No feeds are now allowed to contain animal products.

In view of past practices within the feeds industry, it is important to maintain a questioning attitude in relation to feeds. When buying, check the label on the sack! All ingredients should be listed in a readable form, so that it is quite clear what additives, if any, have been included. The only necessary additives in poultry feeds are mineral and vitamin supplements.

The label shown below is clear and helpful. The feed has no additives, other than vitamins and minerals. The labelling is also comprehensive, with a contact number for the buyer to obtain further information, if needed.

FARMYARD LAYERS PELLETS 1125

Oil%	Prot%	Fib%	Ash%	Moist%
3.2	17.0	6.2	12.5	14.0%

Vit A iu/kg	Vit D_3 iu/kg	Vit E iu/kg	Copper mg/kg
8000	2400	12	23

Methionine: 0.28%
Added copper as cupric sulphate Serial No. 701884
Best before/Vitamins present until End Oct 1996
RATION INGREDIENTS (in descending weight order)
Wheat, Wheatfeed, Sunflower meal, Limestone grit,
60% Maize gluten, Beans, Lucerne,
Hipro soya, Peas, Vegetable fat, Minerals, Vitamins
This is a complete compound feed for laying hens.
For further information please telephone 01245 354455.
Store in a cool dry place.
Weight: See bag/bulk weight declaration.

Source: Marriages Feeds.

Figure 7.3 Example of a compound feed label.

Producers are now required to keep records of the feed ingredients used in their feeds. Many layers' feeds vary in their composition in order to take into account the differing requirements of the following stages: pre-lay, early lay, peak lay, mid-lay and end of lay. Most feed companies will advise on the requirements of a particular situation. Breed companies will also advise on the nutritional requirements of their stock. These change to reflect the needs of newly developed and managed strains. Table 7.1, for example, indicates the daily nutritional requirements of the ISA Brown when in lay.

Tailoring feeds to suit the requirements of various stages of lay makes sense, as long as *all* the factors are taken into consideration. Extra linoleic acid and methionine with high energy feeds, for example, can increase the size of the eggs. This may seem to be a good thing, for large eggs bring the best return. The other side of this is that it may be imposing too much strain on the chickens. Prolapse and vent-tearing problems caused by straining to lay over-size eggs are becoming more common. These problems, in turn, attract the attention of other birds, leading to vent-pecking. Another effect of 'too rich' a diet is that of feather loss out of the normal moulting period. This is linked to

Table 7.1   Daily nutrient requirements of the ISA Brown in lay

Protein		Energy (at 20°C)		
Crude protein	19.5 g/d	Kcals/day – 2775 Kcals (11.6 MJ)		
Lysine	880 mg/d	Levels of production	Cages	Floor
Methionine	430 mg/d	2–10%	250	270
Methione & cystine	760 mg/d	10–30%	260	280
Tryptophan	190 mg/d	30–60%	285	305
Isoleucine	700 mg/d	60–95%	310	330
Threonine	620 mg/d	After peak	320	342
Valine	820 mg/d			

Phosphorus, Calcium, Sodium chloride, Linoleic acid

	Age (weeks)		
	18–28	29–50	50+
Phosphorus[1] g/d	0.44	0.42	0.38
Phosphorus[2] g/d	0.40	0.38	0.34
Calcium g/d	3.65–3.75	3.8–4.2	4.2–4.6
Sodium min mg/d	180	180	180
Chloride min/max mg/d	170/260	170/260	170/260
Linoleic Acid min g/d	1.6	1.4	1.25

[1] Use this level when calcium is supplied as limestone flour
[2] Use this level when 65% of calcium is supplied as granules 2.4 mm

Vitamins (added/kg feed)

A	10,000 IU
D₃	2,000 IU
E	20 mg/kg
K₃	2 mg/kg
B₁	2 mg/kg
B₂	5 mg/kg
B₆	3 mg/kg
B₁₂	0.01 mg/kg
Niacin	25 mg/kg
Pantothenic acid	10 mg/kg
Folic acid	0.5 mg/kg
Choline chloride	500 mg/kg

Trace elements (added/kg feed)

Manganese	60 mg
Zinc	50 mg
Iron	50 mg
Iodine	1 mg
Copper	5 mg
Cobalt	0.2 mg
Selenium	0.15 mg

Source: ISA Brown Newsletter, May 1996.

A range of outside and weather-protected feeders. The large one on the left dispenses feed on demand. *(Solway Feeders)*

two other factors – stress and lack of exercise. Birds in very large houses are subject to inappropriate conditions and they do not get out of their houses for a sufficiently long time to 'burn off' the high energy feeds.

The problem of stress and inactivity leading to feather loss is not new. It has long been known in the battery cage sector. Birds often emerge from there with hardly any feathers at all. It is a relatively new phenomenon in the free-range sector, where it is increasingly appearing in flocks housed in large houses.

Housing and feeding less intensively, and encouraging the birds to get out of the house more, will go a long way to solving the problem. Small poultry keepers and producers with low density flocks do not experience the problem to the same degree. They have long known the value of feeding grain as well as compound feed. They also feed grain outside, to encourage extensive ranging, and they do not subject their birds to the stress of over-large flocks.

Large producers need to think for themselves, rather than rely on computer systems that would have them using

Outside feeder from which hens can feed when they wish to, and which is secure against vermin or wild birds. *(Hengrave)*

compound feeds to the exclusion of all else. They also need to think from the 'outside-in', rather than viewing the house as the centre from which chickens emerge only as a sop to appease welfare considerations. Free-range is not placing an intensive house in the middle of a field and hoping that all the chickens stay in!

Grain

Free-range birds normally consume up to 15 g of grain per bird per day, although this varies considerably depending on weather and other conditions. It provides a balance of nutrients, along with the compound feed ration. In winter, the grain ration can be increased to 20 g per bird per day in order to compensate for the extra energy needed to keep warm. The point has been made earlier that feeding extra grain is cheaper than making more layers' mash available. Suitable grains are wheat and kibbled (chopped) maize.

The recorded results for Clem Shaw's utility flocks of pure breeds on page 7 make interesting reading. Their consumption of grain is considerably higher than it is for most commercial hybrids.

The grain ration should be fed outside, in a different area every day. It can be put on the ground as a scratch feed, but this also provides food for wild birds. If chickens are to take full advantage of their grazing area, they may need incentives to wander further afield. The best way of doing this is to have some grain feeders and drinkers well away from the house.

There are feeders, as referred to earlier, with a displacing mechanism that allows a chicken to dislodge a certain amount of grain on demand. This ensures that the right amount is made available and can be consumed before wild birds realise what is going on. The mechanism is too high for rodents to have access.

There are also feeders with a timing mechanism, allowing the dispersal of grain to be programmed for specific times. These feeders were originally designed for pheasant rearing, but free-range poultry producers have realised their potential for chickens.

Shade and protection should be provided where outdoor feeders and drinkers are placed.

Grit

Most free-ranging birds will pick up small stones and pieces of grit. These are essential for the efficient working of the gizzard for breaking down grain, as mentioned above. It is recommended that insoluble grit should be made available, either in small hoppers or on the ground, at the rate of 20 g per bird per month. This will ensure that reluctant rangers or soils lacking stones will be taken into account. Insoluble grit for chickens is widely available from poultry suppliers.

Coarsely ground oystershell, as referred to earlier, will also ensure that calcium intake is sufficient to provide strong-shelled eggs. This can be placed with the insoluble grit so that the chickens can help themselves.

Water

Chickens, like all living creatures, need fresh, clean water available during the day. This is not surprising when you consider that nearly 70 per cent of a hen's weight is water, and an egg contains around 65 per cent water. Five hybrid birds will drink about 1 litre of water a day in normal conditions, with consumption more than doubling in hot weather. A water shortage which continues for up to 5 hours will cause the birds to eat less and the eggs will be smaller as a result. It may also trigger an egg eating outbreak. Water which is too warm is also unpopular, as the experience related on page 99 indicates.

Large units will have an automatic water supply consisting of a header tank, feeder pipes and drinkers. The latter are usually suspended over the droppings pit area – never over the litter area; otherwise a breeding ground for coccidia is provided! Small units generally rely on manually filled drinkers.

As long as the header tank, pipes and drinkers are kept thoroughly clean, problems

A large outside drinker placed on wooden slats to avoid muddy conditions.

A galvanised steel drinker that can be suspended if necessary.

rarely materialise in summer. Winter can be a different story. Frozen pipes are a real menace and every effort should be made to lag water pipes effectively. An electrical heating tape is an effective way of guarding against frost ravages. It is self-regulating in that it only heats up when the temperature approaches freezing point. It can be wound round the pipes and tap and, once switched on, can be left to come into action if the temperature falls drastically.

In really hot summers there may be a problem with outside drinkers unless they are shaded. The birds are unlikely to use them if the water is hot and will anyway appreciate a shaded area for their own use. It is well worth putting up some kind of temporary and easily moved shading where the drinker is situated. One poultry farm I visited had a large beach umbrella placed for the benefit of the birds. It looked slightly incongruous but the important thing was that it worked!

Freedom Food feed requirements

If eggs are being sold under the Freedom Food marketing initiative, the following must be adhered to, in addition to the grass ranging and housing requirements referred to in earlier chapters.

- There must be a wholesome and appropriate diet, fed every day, in sufficient quantity to maintain good health and to satisfy nutritional needs.

- Food and water must not be in a contaminated or stale condition.
- Birds should not have to travel more than 8 metres in a house to reach food and water.
- There must be an adequate supply of fresh drinking water, with continuous access.
- Provision must be made for supplying water in freezing conditions.
- A minimum feeding space of 5 cm per hen must be allowed in a linear track system, or 4 cm of circular feeding space.

Organic feed requirements

The feed requirements for those producing organic eggs or table birds are understandably more stringent. Reference has already been made to the Soil Association Organic standards for housing and grass ranging. The following apply to feeds:

- Accurate and comprehensive records must be kept of all feedstuffs.
- Poultry diets must contain a minimum of 65 per cent cereals or cereal by-products.
- All poultry must have access to insoluble grit.
- The system should be planned to provide 100 per cent of the diet from foodstuffs produced in accordance with organic standards. However, in cases where this is not immediately possible, the organic diet permits 20 per cent of bought-in approved feedstuffs from non-organic sources until August 2005.
- Prohibited are yolk colourants, in-feed medications or any other feed additives.

There are specialist suppliers of organic feeds and constituent feedstuffs in most areas of the country, although these are generally more expensive than those from non-organic sources.

Mixing your own feed

It is not generally feasible to try to mix a ration for a flock, unless there is a specific reason, such as in the running of an organic flock. Even so, the cost of milling equipment can be high, unless small-scale, traditional tools are used. These include a mill with kibbling plates for chopping rather than crushing large seeds. A mixing facility will also be required.

Table 7.2 gives a suitable mixture for laying birds, if all the ingredients are available. The problem is not so much finding a supplier, as in getting relatively small quantities supplied. Ingredients need to be stored adequately and will obviously go off after a certain period. A local feed mill may be prepared to make up relatively small batches. Vitamin and mineral supplements are available from specialist companies. Many of these exhibit at the annual Pig & Poultry Fair at the National Agricultural Centre in Warwickshire in May.

Table 7.2   Home-made layers' feed of 17% protein (with fishmeal)

	%
Wheat	60
Grassmeal	15
Full fat soya	15
Fishmeal	5
Molasses	2.5
Vitamin & mineral supplement	1.5
Di-calcium phosphate	1.0
	100%

Not everyone will wish to use fishmeal, so maize, peas and field beans can be used as an alternative. A small hand mill can be used to kibble or crush these; small, electrically powered ones are also available. An additional grain ration of wheat is required for adult birds, as well as access to grit.

Kitchen and garden scraps

If the eggs are being sold, all scraps should be avoided. Where just a few birds are being kept for the household, it is commonsense to give leftovers to the birds, as long as they are fresh and there is no meat or excess salt. Scraps can be a useful source of supplementary food, as was realised in wartime conditions. It should be remembered, however, that they should be an occasional addition rather than the main nutrients. Digestive upsets are far more likely if scraps are included in the diet.

Kitchen garden greens are appreciated, particularly lettuce, cabbage and other brassica plants. If these are suspended in the run they will provide interest as well as a feed supplement. This is particularly important in a small run where the chickens may be prone to boredom, a situation which can lead to practices such as feather pecking or egg eating.

Alternative food sources

A traditional practice to overcome the lack of green food in the winter is to use sprouting grain. Any grains can be sprouted, but wheat and oats are most commonly treated in this way. There are several ways in which it can be done, but the basic technique is to soak the grain in warm water for twenty-four hours, drain off the water, but continue to sprinkle the grain with water every morning and evening. When the grains begin to sprout, spread them out on a clean surface and continue sprinkling until the sprouts are about 3 cm long. They are now ready for feeding, either as they are or mixed with other feeds.

Some research has been carried out in Australia and Britain on the use of agricultural lupins (not toxic garden lupins) as a potential source of poultry feed.[5] No doubt there will be further research into new sources of vegetable proteins for poultry feeds now that animal proteins have been discontinued.

Finally, an interesting system that I saw used in the Dordogne area of France is the old practice of 'shucking'. Crops are grown specially for the chickens, which are allowed to forage them where they stand. The main crops used are sunflowers, maize and kale.

The birds are allowed access to the area when the plants are ready for picking. They peck out the seeds from the maize cobs first, and when the sunflowers bend to shed their seeds, these are then taken. In my experience, however, chickens will not take sunflower seeds unless they are hulled and kibbled. Perhaps the Dordogne chickens were tough. One thing is certain; they were not beak-trimmed!

References

1. Soil Association, *Organic Symbol Standards*, 1996.
2. Ian Burrell, 'Ministers Fear Maize Breakout', *The Independent*, 4 December 1996.
3. Dr Norman B. Magruder, *Calcium Distribution and Environmental Studies*, Cargill Inc, 1965.
4. Hull & Scott, *Relation of Dietary Calcium Source and Level to Over-consumption of Feed*, Oyota, 1969.
5. Andrew Walker & Heddwyn Owen, 'Vegetable Protein Sources for Laying Hen Feeds', *Poultry Progress*, ADAS, May 1996.

CHAPTER 8 Poultry Management

Sound poultry husbandry should include due consideration of their behaviour.

A.H. Sykes, 1971

In Chapter 2 we looked at the behavioural characteristics of chickens. Having a knowledge of these enables us to be properly prepared for looking after them. Having a 'good eye' to spot potential problems is an asset. Where this is supplemented by adequate training, it is even better. Anyone thinking of starting a free-range farm should only consider doing so after attending a course on free-range management. When knowledge and a house are in place, then the introduction of pullets can be considered.

Buying pullets

The best time to buy laying stock is at the age of 16 to 18 weeks when they are at the 'point of lay' (POL) period. Laying normally starts from 20 to 21 weeks and continues for between 52 and 60 weeks. In other words, the hens will be around 72+ weeks of age when they come to the end of the laying period. Acquiring the young pullets a few weeks before commencement of lay gives them time to settle down and to become accustomed to their new environment. Mortality levels are also considerably less than they would be with chicks.

Those who are new to poultry and who may be keeping them in a domestic flock are also well advised to start with point of lay pullets.

If hybrids are purchased, it is important that they are floor-reared with the ability to perch. There are specialist rearers of hybrids for the free-range sector who ensure that their birds have been reared on a long step-down lighting programme to reduce competitive stress and give the young birds a longer period to feed. This effectively delays maturity by a couple of weeks and does away with small eggs at the beginning of lay.

Commercial point of lay hybrids bought from specialist rearers will have been inoculated against Newcastle disease, Marek's disease, gumboro disease, epidemic tremors, infectious bronchitis and egg drop syndrome. Check with your local veterinary surgeon or Ministry of Agriculture poultry specialist, in case there are any other diseases prevalent in the area. Commercial rearers are also required to test their stock for salmonella on a regular basis, and will usually provide the buyer with a record of this in relation to the new pullets.

Most rearers beak-trim hybrid chicks as a matter of course. If untrimmed hybrids are preferred, you will need to arrange this with the rearer in advance.

If pure breeds are bought, they may not have been tested for salmonella and vaccinated against the diseases referred to above, unless they come from a large breeder.

Introduction to the house

Once the house is in position and equipped,

it is ready for occupation. Feed and water should be available. The ideal time to introduce chickens to their new house is in the evening, although this may not always be a practical proposition. There should be enough people present to cope with the number of birds. Even with just a few birds, there should be two people to handle them.

After travelling from the supplier, the birds will be under a certain element of stress, so calm, unhurried behaviour on the part of the handlers is essential. Gentle handling is also necessary. The correct procedure is to slip one hand under the bird and hold it firmly by *both* legs, with two fingers between the legs, while the other hand is placed on the back to restrain the wings. When lifted up in this way, and held close to the handler's body, the bird is properly supported, without risk of damage to the wings. Ideally, one bird at a time should be carried.

The pullets should be placed gently in the house. Talking to them quietly has a calming effect. Putting some on the perches and others near the drinkers will help to train the rest. Many newly housed birds need to be shown where the food and water supplies are positioned. If they are still fairly young, they will be on a growers' ration. Just before point of lay, they can be switched to a layers' ration. As with any change, this should be gradual rather than abrupt, but from an early age, they should be encouraged to go outside and to have a grain ration.

It is a good idea to close off the nest boxes where growing pullets are concerned. This will prevent them going to sleep in them instead of on the perches. Once the birds are used to their house, and to the perches, the nest boxes can be opened as point of lay approaches. Some very large houses have nest boxes that are closed every night and opened early every morning.

When all the chickens are inside, leave them to get used to their new home, but keep an eye on them, from time to time, to make sure that there are no problems of panic or

stress. On a commercial scale, the pullets should continue to follow the lighting regime they have had at the rearers. It is important that they do not have too much extra light; otherwise they may start laying before they are sufficiently grown. Further details are given in the section on lighting later in this chapter.

If a small number of birds are introduced to a domestic house in late evening, they can be left overnight. If it is day-time, and the feed and water are outside in a run, they can be let out into the run after an hour's confinement. This allows time for the perching/nesting association to imprint itself on the birds before they go out. Once they are released, they will soon find their way out into the run, and will then remember how to get back inside. Putting them outside first can cause problems, with some taking a long time to learn how to go up the ramp into the house. Once 'home' is imprinted, however, every effort should be made to persuade them to range outside. Reference has been made in Chapter 7 to the problem of birds in very large houses not getting enough exercise to cope with high energy feeds.

Daily routine

1. Check the outside perimeter fence in the morning to make sure that it is secure.
2. If the fence is electric, use a neon tester or voltmeter to assess the current.
3. If everything is in order, it is time to let out the birds.

Open the pop-holes. Chickens like routine and will expect to be let out at the same time each day, although some take much longer than others to exit. You can start off as you mean to go on and make opening time convenient for you! This is a good moment to check the flock, looking out for any suspicious signs such as feather pecking. A subdued, hunched-up bird or any unusual behaviour should be investigated straight away. The earlier that problems can be sorted out, the better. A list of common problems and diseases is given in Chapter 14.

Check manual drinkers and feeders. If soiled, clean them out and replenish with fresh water and feed. On a large scale operation, check the rate and flow of water in automatically delivered systems. Adjust the ventilation in the house by opening vents or windows as necessary, and check the timer controlling the lighting.

Ensure that there is grit and crushed oystershell available where the birds can help themselves. This can be kept topped up on a regular basis. Where there is a scratching area adjacent to the house, check that it is in good condition. Regular raking gets rid of surface droppings and also aerates the litter. One of the best ways of dealing with pathogens is to expose them to an increased level of oxygen. The aim with any litter is to maintain it in a friable condition.

A visual check of the pasture where the birds are ranging should also be carried out daily, to determine whether it is necessary to move them onto fresh ground. If, for example, bare or muddy patches are appearing, a new area should be made available so that the first can be rested. If the grass is getting high it should be topped, particularly in the area by the electric netting; otherwise shorting of the current may result. Long, coarse grass is disliked by the birds; they cannot graze on it and if it is wet, it makes their bottom feathers wet and muddy. This, in turn, makes the eggs dirty.

Collect the eggs as often as possible, and store in cool conditions. Commercial flocks will have the eggs collected several times a day and taken to the egg store. An egg trolley to go up and down the collecting passage is useful. At the same time, check for the presence of any ground-laid eggs and remove any litter which looks as if it has been designated a nest.

Any eggs that are cracked or soiled should be used or discarded immediately. No cracked or washed eggs can be sold as free-range eggs. Record numbers of eggs and

any other data that may be required. (See page 156 for details on keeping records.)

Check that the wood shavings or other nest box liner is clean, and replenish as necessary.

During the day, carry out occasional checks on the flock to make sure that there are no problems. The outside water supply should also be checked and given shade protection if it is particularly sunny. In the afternoon, give the birds a ration of grain, either on the ground outside as a scratch feed or in a grain feeder away from the house as an incentive to wider ranging. It is a good idea to vary the place every day, so that undue scratching is not confined to one area.

As it gets dark, the birds will go back to the house of their own accord, for the perching instinct is strong.

It is not usually a problem to get the birds back, although one or two individuals may prove difficult, particularly if there are trees on the site. In summer, when the weather is warm, some birds may want to revert to their ancestry and perch on the branches instead of going into the house. If the perimeter fence is secure against foxes there is usually no harm in this, although it may encourage a tendency to lay eggs outside.

Once the chickens have returned to the house in the evening, the pop-holes can be shut and the house secured for the night.

Set any security or alarm systems before calling it a day.

Periodic routine

Small flock houses are periodically moved to a new area of grass. If it is a solid-floored one, remove the droppings board or plastic layer and add the droppings to the compost heap. Brush out the interior of the house and replenish the nest box shavings as necessary.

Larger moveable houses will not need to be moved as often. The frequency depends on the flock density in relation to the ground available. It may be anything from once a month to once a laying season. The type and condition of the paddock is also relevant. When the house is moved, rake the area underneath in order to disperse the droppings. If necessary, lime can be sprinkled on the ground while it is being rested.

Birds in static houses will also need to have access to fresh pasture, but again, rotation depends on stocking levels and grass condition. Further information is given in Chapter 6.

Check on feed stores and buy in as necessary.

Commercial producers will need to buy in pullets on a regular basis so that they have several flocks of differing ages. This ensures that there is no drop in overall egg production. As one flock moults, or is culled, its production is added to, or replaced, by others.

New birds must be kept entirely separate from older ones so that the possibility of disease transference is minimised. In a small domestic flock this may not be possible for more than a few days, but as a general principle it is worth following.

Lighting

Day length has an important bearing on the egg laying cycle. The egg laying mechanism is controlled by hormones that are produced by the action of the pituitary gland in the brain. This, in turn, responds to the amount of light which falls on the bird's head. In autumn, as less daylight is available, the pituitary gland is influenced in such a way as to reduce hormonal action and, ultimately, egg production. The number of eggs is gradually reduced until laying may cease altogether. Poultry keepers can, to a certain extent, mitigate against this by having March or April hatched chicks to take over the winter egg production, but, even so, there is a drastic reduction in the numbers of eggs that are laid. This explains why, traditionally,

Light falls on the hen's head, stimulating the pituitary body in the brain to produce the hormone pituitrin. This is transported in the blood to the ovaries which are stimulated to produce eggs.

Figure 8.1 Effect of light on egg laying mechanism.

there was such a heavy reliance on preserving the summer eggs in a preserving medium such as waterglass (sodium silicate solution).

The effect of light on the egg laying system is not a recent discovery. It was known long before electricity was generally available. Large windows were used in some houses. Traditionally the ceiling and interior walls of a house were painted white in order to make as much reflected light available as possible. Sometimes, a mirror was placed inside for the same purpose. (The areas near the nest boxes were left dark, however, to discourage egg eating.) Paraffin lamps were also used, but these are no longer recommended because of the risks of fire and asphyxiation.

If a bird is given 15 to 16 hours of light a day, she will continue to lay, because her pituitary gland receives information to the effect that it is still 'summer'.

In small houses, it is normally necessary to provide extra light only from autumn onwards through the winter. Large static houses, which are much darker because of their physical size, may need it all year, with the amount varying according to circumstance.

The provision of light needs to be seen in two ways – *day length* and *light intensity*. Both these aspects play an important role and it is necessary to be able to distinguish between them. Day length is, quite simply, the number of hours in which light is available. The longest day is 21 June, when there are 17 hours of light. As this maximum declines, from July onwards, artificial light must be made available to make up the difference, so that birds which are in lay do not have their day length shortened. Domestic poultry keepers would normally start to give light when the birds had finished re-feathering after the moult, in autumn.

Light intensity may be regarded as the strength or degree of brightness of light available. The unit of measurement is a lux, and light intensity is measured with a light

Date	Sunrise	Sunset	Natural daylight hours	Date	Sunrise	Sunset	Natural daylight hours
Jan.				Jul.			
7	08.05	16.09	08.04	7	03.53	20.18	16.25
14	08.01	16.18	08.17	14	04.01	20.11	16.10
21	07.55	16.29	08.34	21	04.09	20.04	15.55
28	07.46	16.42	08.56	28	04.19	19.54	15.35
Feb.				Aug.			
4	07.36	16.54	09.18	4	04.29	19.43	15.14
11	07.24	17.07	09.43	11	04.40	19.30	14.50
18	07.10	17.20	10.10	18	04.51	19.17	14.26
25	06.56	17.32	10.36	25	05.02	19.02	14.00
Mar.				Sep.			
3	06.41	17.45	11.04	1	05.13	18.47	13.34
10	06.26	17.57	11.31	8	05.24	18.31	13.07
17	06.10	18.09	11.59	15	05.36	18.15	12.39
24	05.54	18.21	12.27	22	05.47	17.59	12.12
31	05.38	18.33	12.55	29	05.58	17.43	11.45
Apr.				Oct.			
7	05.23	18.45	13.22	6	06.09	17.27	11.18
14	05.07	18.56	13.49	13	06.21	17.11	10.50
21	04.52	19.08	14.16	20	06.33	16.57	10.24
28	04.38	19.19	14.41	27	06.46	16.43	09.57
May				Nov.			
5	04.25	19.31	15.06	3	06.58	16.30	09.32
12	04.13	19.42	15.29	10	07.10	16.19	09.09
19	04.04	19.52	15.48	17	07.22	16.08	08.46
26	03.55	20.01	16.06	24	07.34	16.01	08.27
Jun.				Dec.			
2	03.49	20.09	16.20	1	07.46	15.55	08.08
9	03.44	20.16	16.32	8	07.55	15.52	07.57
16	03.44	20.19	16.35	15	08.02	15.52	07.50
23	03.44	20.22	16.38	22	08.07	15.54	07.47
30	03.48	20.20	16.32	29	08.08	15.59	07.51

Figure 8.2 Daylight hours (Greenwich Mean Time).

meter. The table below gives an indication of lux measurements.

Table 8.1 Typical illuminances

	Lux
Bright sunlight	80,000
Overcast day outside	5,000
'Bad light stops play'	1,000
Modern office or factory	500
Side road lighting	5
Moonlight	0.2
Clear starlit night	0.02

Source: Essentials of Farm Lighting, Electricity Council, 1987.

The best type of house is one that has natural daylight coming in. Many large houses, however, are extremely dim inside. In these, there should be a minimum of 10 lux; otherwise the birds may have difficulty in finding their way about. This is also the minimum allowed under Freedom Food requirements. A lighting system with a dimmer attachment can be used to increase or decrease the intensity, as needed.

It is a fact of life that in large houses there is an increased risk of aggression because of the numbers of birds involved. Where aggression breaks out, it may be necessary to dim the light slightly. Getting the balance right can be difficult, for if the decrease is too much, or too sudden, it can cause a reduction in laying. If it is too great, it may encourage pecking, egg eating and other vices. The important thing to remember is that any change should be gradual.

Lighting can be provided in one of several ways, depending upon the size of the house; 40 watt tungsten bulbs or 6–8 watt fluorescent bulbs or tubes are satisfactory, with the light sources placed 10 m apart. One light source is sufficient for up to 100 birds.

On a small scale, an ordinary 25 watt bulb is satisfactory for an outbuilding, while a portable system based on a 12 volt car bulb and battery suffices where mains electricity is not available. A rechargeable Nicad battery is also suitable.

Whatever system is used it is important to incorporate a time switch so that the amount of lighting can be controlled. Time switches are available for 12 volt systems, as well as for mains operated ones; it is important not to confuse the two.

A dimming facility is also important to warn birds that the lights are soon to be extinguished, giving them time to find their way to the perches.

These are the golden rules for lighting:

- Do not provide extra light too early, before point of lay pullets have grown adequately, or they will lay early and the eggs will be small. They may also have problems of coping with large eggs later if they are young.
- Increase the period of light gradually until the maximum of 15 to 16 hours is reached.
- Do not allow the day length to shorten once the birds are laying.

Free-range birds must be adequately grown before they can cope with the demands of outdoor ranging as well as egg laying. Figure 8.3 gives the lighting recommended by Grassington Rangers, who specialise in rearing pullets for the free-range sector. Their pullets are reared on a long step-down lighting programme, which delays sexual maturity by 10 to 14 days and eliminates early, commercially worthless eggs. The flock is housed at 17 weeks with an initial one hour extra light being given to trigger laying at around 21 weeks.

For flocks in small houses, no artificial light at all need be considered until autumn. If the birds have gone through the moult and are re-feathered by late autumn, give them artificial light, initially an hour, then gradually increasing by half an hour a week until a total of 16 hours (natural and artificial) is given. More than 16 hours outside the normal mid-summer period should not be given on humanitarian grounds.

Month	Natural Day Lengths (hours)	Wk 18	19	20	21	22	23	24	25	26	27	28	29	30
JAN	8–9	9	10	11	12	13	13	14	14	15	16	———————→		
FEB	9–10.5	9	10	11	12	13	14	14	15	15	16	———————→		
MAR	11–13	9	10	11	12	13	14	15	15	15	16	———————→		
APR	13–14.5	9	10	11	12	13	14	15	16	———————————→				
MAY	15–16	10	12	14	14	15	15	15	16	———————————→				
JUNE	16–16.5	10	12	14	15	15	15	16	——————————————→					
JULY	16.5–15.5	10	12	14	15	15	15	16	16	———————————→				
AUG	15–14	10	12	14	15	15	15	16	16	———————————→				
SEPT	13.5–11.5	10	12	13	14	15	15	15	15	16	———————→			
OCT	11–10	9	10	11	12	13	14-	15	15	15	16	———————→		
NOV	9.5–8.5	9	10	11	12	13	14	14	15	15	16	———————→		
DEC	8–9	9	10	11	12	13	13	14	14	15	16	———————→		

Figure 8.3   Guide to artificial daylight for pullets housed at 17 weeks. *(Grassington Rangers)*

Coming into lay

The behaviour of the pullet about to lay her first egg is unmistakeable. She will make a slightly complaining, almost continuous crooning sound and may go to and fro from the nest box several times before she eventually settles down. When it arrives, the first egg may be quite small and may even be a 'wind' egg – one without a yolk. This is quite common, for it may take the egg laying mechanism a few days to get into its stride. Once the egg is laid, the pullet will emerge from the house with the familiar cackling sound which announces the fact to the world.

The period of laying is also indicated by the fact that the pullet's pelvic bones gradually move further apart. At first this distance will be about two fingers' width, as indicated in the photograph on page 97. As the hen gets further into the laying period, the space becomes bigger, so that it measures three or even four fingers' width.

It is important to ensure that the birds lay in the nest boxes, not on the floor. The use of 'pot' (pottery) eggs will encourage them, as well as the provision of a nest box liner. The nest boxes should be in a darker area than the rest of the house. Where rollaway nest boxes are not used, a plastic curtain can be placed across the opening. This not only provides a quiet, darkened area, but also helps to prevent egg eating.

If there are any floor-laid eggs, they should be removed as quickly as possible.

Moulting

Moulting is the natural process of losing old feathers and replacing them with new ones. It takes place once a year. Autumn- and winter-hatched birds will moult between July and August. Those hatched after March will normally continue through to October or November before moulting starts. Moulting usually lasts a few weeks and egg production declines, and may even cease, while it is going on. The laying hybrid which has been bred for maximum production is less likely to

This pullet has been laying for several weeks. The space between the pelvic bones is two fingers' width. As she gets further into lay, the width will increase.

cease laying all together than the older, pure-bred bird. Bear in mind, however, what was said in Chapter 7 about how the combination of stress, 'rich' feeds and lack of exercise can bring about unnatural moulting.

The first indication of moulting is a loss of feathers around the neck, followed by a gradual dropping of feathers from the abdomen, back, breast and tail. Occasionally, a bird may be affected by 'drop' moulting; the feathers tend to fall more or less all at the same time. Moulting puts a considerable strain on the bird's constitution. Chickens which drop moult will need extra care because they are weakened and therefore more prone to illness. Make sure that they are adequately fed. Protein, vitamin and mineral requirements are particularly important at this stage. Traditionally, chickens were also given poultry spice, which is a supplement rich in minerals and vitamins which helps to ensure speedy re-feathering. It is available from poultry suppliers and is often used for show birds.

With several flocks of different ages the moulting periods occur at different times so that any decline in the number of eggs is minimised when the overall number of birds is considered.

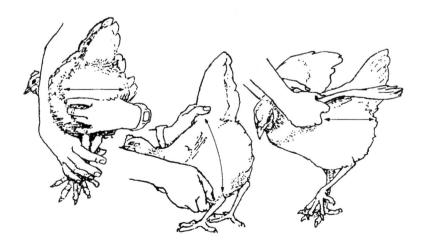

Figure 8.4   Judging a good layer. The areas indicated are more generous than in a poor layer.

Figure 8.5   The moulting process: the numbers indicate the sequence of feather loss.

It used to be the custom to 'force moult' laying flocks so that they had moulted and re-feathered before the autumn. This was to ensure that they started laying again before the daylight hours dwindled too much. The advent of artificial lighting and readily available point of lay pullets at different times of the year makes such practices largely redundant.

Traditional force moulting, by keeping birds in a hot, dark house, with access only to oats and water for a couple of weeks, is not recommended on humanitarian grounds. What is slightly more acceptable, if a flock is to finish the moult at the same time, is to keep them in their house with adequate ventilation and normal rations, but to provide a maximum of 8 hours of light. There is far less stress on the birds in this way, and egg supplies can be continued by another flock of a different age. The moult usually starts after 2 weeks of confinement. It is important to remember that they should be able to exercise, move about and scratch in clean litter during this time. While the birds

are so confined, any eggs which are produced cannot be legally sold as free-range ones. Once the birds are refeathered, the lighting pattern is stepped up, with a gradual increase to a maximum of 16 hours.

I do not recommend this practice, and indeed those who are registered with Freedom Food are not allowed to follow it. I have always allowed my birds to moult naturally, without confining them or trying to make them all moult at the same time. Being free to moult and then recover gradually, in their own time, is less stressful to the birds. It also gives them a much-needed break from egg production. They will then be well prepared for winter laying. The point was made earlier in the book that although chickens are highly developed layers, they do not have to be treated like machines.

Summer management

During the moulting period a careful watch should be kept for any evidence of stress, feather pecking or parasites. Lice, mites and ticks can appear at any time, but are particularly prevalent in warm weather. Lice can be seen as small, grey creatures moving along the skin when the feathers are parted. Ticks are bigger and are embedded in the skin. They should not be pulled out, for the head will remain behind, causing an abscess to form. Spray on or dust the appropriate proprietary product and leave the dead ticks to drop off.

Red mites hide in crevices in the house during the day, and come out to feed on the hens while they are perching at night. It is necessary to treat the birds, the house and nest boxes, as well as the dustbathing area against them. Proprietary products, both powder and spray, are available from licensed suppliers and vets. An example is a product called Ban-Mite. There is also a product that is added to the drinking water, but I do not recommend this for egg layers, although it may be appropriate for show birds which are not associated with the human food chain.

Mites can also get under the scales of the legs, pushing them up and forming white encrustations. Scaly leg can be dealt with by painting the legs with liquid soap and disinfectant, and rubbing gently with an old toothbrush to dislodge the crusts. Pulling them off will only cause the skin to bleed. Again, a proprietary product will kill the mites. It needs to be administered several times before the legs are clear.

Free-ranging birds are also susceptible to worms, not only from over-use of the ground, but also via pheasants and wild birds. If this happens, they will need to be treated with a suitable vermifuge. Flubenvet is suitable for all poultry worms – gastro-intestinal and respiratory tract nematodes and cestodes – and is available from licensed suppliers. It is mixed with the feed. There is a withdrawal period of seven days after the last dosage, so any eggs laid in this period should be discarded. It makes sense, therefore, to administer it when the birds have ceased laying, during the moult. If it is necessary to administer it at other times, the eggs must be discarded.

In hot weather, chickens can succumb to heat exhaustion. Signs are the wings being held out slightly from the body, and the beak left open. It is important that shade protection is provided for them and their outdoor drinker. Their consumption of water can double to what it normally is (5 litres a day for 25 birds).

The temperature of the water is also important, as one Israeli farmer discovered. He noticed that his birds were drinking less, despite the fact that drinking is a way of maintaining the normal body temperature of 39°C. It was only when the water was cooled that they began to drink. The farmer devised a method of maintaining the same water temperature all year round, and the Israeli Ministry of Agriculture tested its effect. They found that deaths from heat exhaustion dropped from 12 per cent to 2 per cent, while production rose by 10 per cent. The egg laying period was prolonged by a week and the number of broken eggs was halved.[1]

Winter management

Winter can be a time of hardship for outdoor birds, particularly those in small houses. A house with a small run needs to be in a protected area, away from the wind. The point has been made earlier that it may be more appropriate to have the birds in different winter accommodation. If this is so, it is an excellent time to give the summer house a thorough clean. If the timber needs renewing, or if there has been a problem with parasites such as red mite, the house, perch and nest boxes can be treated.

I housed my layers in a converted stable with light in winter. They had access via a

One way of keeping cool is to hold out the wings slightly. Heat is also lost via the comb and wattles. In severe heat the beak is held open.

pop-hole into a south-facing yard which, in turn, led out to the orchard. There was a gate into the orchard, so they could be confined to the yard, if necessary. When there was heavy snow, they had access to the yard only.

Where chickens are in a house and run, it may be a good idea to put up a windbreak along the sides of the run. Depending on the prevailing winds, this may only be required on one side. Close-meshed horticultural plastic is suitable. If the run does not have a roof, it may also be a good idea to partially cover it, so that there is outdoor protection as well as an open space.

One of the biggest problems in winter is that of freezing water in the drinker. This needs regular checking and replacing. One of the most effective ways of dealing with this is to use electrically heated tape, available from electrical and greenhouse suppliers. This can be used with a thermostat so that heating only occurs when the temperature drops to freezing point. It can be used to provide protection for taps and pipes, as well as underneath the drinkers themselves. It goes without saying that such a system should have a voltage regulator and be checked by a competent electrician.

Other remedies that have been suggested are adding a little glycerine to the water or placing a tennis ball in an open container of water. Neither of these is really effective. Glycerine can get into the nasal passages of the birds, and in cold areas of the country, the water still freezes solid. The idea of the ball is that the birds push it down and in doing so, break the ice and gain access to the water. This is fine if the ice has barely formed. There are only two ways to overcome the problem: either change the water very frequently, or use a heated tape system.

In really bitter weather, the chickens' combs are vulnerable to frost-bite. One poultry keeper in Scotland makes a point of applying Vaseline to his birds' combs so that they have some protection.

In winter, the chickens will need extra grain to enable them to cope with the weather. Details of this are in the chapter on Feeding.

If chickens have protection, light, food and water, they give a lot in return. A fresh egg in winter is even more welcome than one in the summer. As far as commercial producers are concerned, there should be no drop in production as long as extra feeding to keep warm is combined with freedom from stress and active exercise.

Coping with predators and vermin

The main enemy of the free-range producer is undoubtedly the fox. He is a perennial problem, particularly during autumn and winter when other sources of food are scarce. Normally, foxes hunt at night, starting their rounds at dusk, but in winter they may make much earlier visits while it is still light and before the chickens have been confined to their house. Chapter 6 gives details of fencing that will keep them out.

On a small scale, it may be difficult to provide adequate perimeter fencing. Over the years many ingenious ideas have been tried to deal with the fox, including a trapdoor device which will support the weight of a chicken but not that of a fox. Creosote painted along the ground at the enclosure boundaries is also said to be effective. But perhaps the most ingenious suggestion, for those living near a zoo, is the claim that lion or tiger dung placed strategically around the boundary of the run will send the fox fleeing for his life! Having a dog on the premises also helps. There are suppliers of fox traps which lure the fox into a metal cage. It does not harm him, but it does present the problem of what to do with him once caught. A few years ago, one irresponsible local authority rounded up foxes that had moved into their urban areas, and then transported them to mid-Wales to release them. I had an anguished telephone call from a poultry keeper there who suddenly found three totally disorientated foxes in her yard in the middle of the day!

The local hunt, despite protestations to the contrary, is quite inadequate at controlling the fox population. I once had the hunt rampaging arrogantly through my field, frightening the chickens and disturbing the beehives. Fortunately my Siamese cats were indoors. A few days later, I had a visit from the fox. On balance, he did less damage.

It is not possible to exclude the rat, so regular control to keep down numbers is essential. Watch out for tell-tale signs such as droppings and gnawed areas of woodwork. Rats are close to human beings wherever they live. The sewage system of underground passages has provided them with an ideal breeding ground, and there is nowhere which is free of them. If poultry feed is left lying around, it will attract rats. It must be said, however, that poultry feeds which are based on plant proteins have a far less pungent smell than the old feeds which had animal and blood proteins in them. All feedstuffs should be stored in rodent-proof buildings. Mice can also be a problem, although not to the same extent.

Rats are particularly dangerous because they carry Weil's disease, a condition which is potentially lethal in man. It is transmitted via rat urine. Rats also carry pasteurella and salmonella.

Make sure that rats cannot get into a small run or the house. It may be a good idea to have wire mesh on the ground so that they are unable to dig their way in. If the unit is moved regularly, this may not be necessary.

If rats are seen, they should be dealt with immediately. Householders can contact their local authority who will provide the services of a rat catcher, free of charge. On commercial premises, the owner must pay for this. The Poultry Laying Flocks (Collection and Handling of Eggs and Control of Vermin) Order, 1989 also makes it a mandatory requirement for egg producers to prevent vermin infestation of poultry houses and egg stores. Producers normally have a contract with a specialist firm whose representatives call on a regular basis. Such a contract can be expensive.

Poison needs to be placed carefully so that only the rats have access to it. It should be quite inaccessible to the chickens, wild birds, domestic pets and, of course, children. Companies which sell rodent control products provide excellent advice and guidance on the placing of rat poison, and their suggestions should be followed.

Mink, which are a problem in some areas, are normally found near lakes and rivers. There is really only one solution and that is trapping. The traps need to be specific to mink so that other wildlife and domestic pets are not endangered. The Ministry of Agriculture will advise on this.

Replacing and culling the flock

Free-range producers are normally advised to cull flocks at around 74 weeks, or to moult them at around 60 weeks for another laying cycle. In other words, the birds are kept through their first laying period, the moult and a subsequent period of lay. Once that comes to an end there is an overall decline in the number of eggs. What needs to be taken into consideration is that, with older birds, there is likely to be a higher incidence of misshapen eggs, and birds with oviduct prolapse problems and broodiness. They should certainly have a feed which is toned down so that linoleic acid and high energy elements are reduced. Many small producers keep their layers for a second year.

On a large scale, cull birds are normally sold to contractors who come to catch and crate the birds to be taken for slaughter. Franchisers who supplied the birds will normally have a 'buy-back' scheme. The Humane Slaughter Association is concerned that catching and slaughtering should be done as humanely as possible and offers excellent advice in its booklet *Practical Slaughter of Poultry*. Its content is applicable to anyone who has to catch and slaughter birds, on any scale, and it should be on the shelf of every poultry keeper.

In large houses, the lighting should be dimmed to avoid panic and smothering. It is important to ensure that cull birds are not deprived of food and water. Erecting small pens will enable them to be confined before catching. Modules made of galvanised steel and mesh, with nylon wheels for quiet use are available. A calm, quiet approach will prevent stress and damage to the birds. They should be caught by grasping both legs and held in such a way that they are properly supported.

On a domestic scale, many birds are kept till the end of their natural lives because they are often regarded as family pets.

There are times, of course, when a hen has to be put down, either because she is old and infirm, or because she has been in an accident. It is essential that this is carried out by someone who is competent to do it, without causing any distress to the bird. There is a requirement in the Welfare of Animals (Slaughter or Killing) Regulations, 1995, that slaughtering should be carried out by those with the knowledge and skill necessary to perform the task humanely and efficiently.

Neck dislocation is a legal method of killing a hen, but is regarded as an emergency method only. It is not recommended for the routine slaughter of birds. For this, electrical stunning followed by neck cutting is more appropriate.

Where neck dislocation must be carried out, it is a matter of simultaneously twisting and pulling the neck, so that it breaks quickly. Do the actual killing out of sight of the rest of the birds. This may seem unnecessarily considerate, for chickens are not very intelligent, but while there is the slightest possibility of distress being caused,

it is worth taking a little trouble to avoid it. Once the killing has taken place, the chicken will flutter its wings for a short time. This is normal; it is a nervous reaction that always follows death.

Cleaning the house

When a flock is replaced or, in the case of a small one, goes to different winter accommodation, the original house needs thorough cleaning and disinfecting.

On a large scale, this will involve removing manure from the droppings pit in a slatted floor house, or litter from a deep litter area. Contractors will do this, with the cost of the manure as a fertiliser often being deducted from the charge. Alternatively, the producer may wish to do it himself and compost the droppings for subsequent bagging and sale. Equipment such as a Bobcat skid-steer loader or a tractor with a front-end loader will be needed. On a small scale, a stiff brush will usually suffice, with the droppings going on the compost heap for subsequent use in the garden.

Disinfecting is a good idea to ensure that there are no pathogens likely to be transferred to a new flock. On a large scale, this can be applied in conjunction with a pressure cleaner.

Small timber houses can be partially dismantled and treated with disinfectant. Where there is an infestation of red mites, spray with a proprietary product for the purpose.

References

1. Division of Poultry Science, Agricultural Research Organisation, The Volcani Centre, Israel.

CHAPTER 9 **Eggs**

We must realise that it is the consumer who defines quality, not the producer.

John Riley, ADAS Poultry Adviser, 17 May 1989

The egg starts as a small, unfertilised ovum attached to a yolk in the ovary of the bird. It is released from the ovary and enters the top of the oviduct where it acquires a coating of albumen, or egg white. Further down the oviduct it is covered by two membranes and, finally, a shell is secreted around it by a special gland. It emerges from the vent, between 24 and 36 hours after leaving the ovary. When first laid, it is damp, but it quickly dries and develops a protective bloom, which helps to keep out dirt.

In 1859 Mrs Beeton extolled the virtues of the egg as a 'delicate food, particularly when new laid'. Today's EU description of such an egg as 'class A' may not have literary merit, but it means the same thing. All class A eggs are fresh, and these are the ones that free-range producers will be aiming for, whether they have commercial or domestic chickens.

Egg collection and storage

Eggs should be collected as frequently as possible, without disturbing the chickens. Unless rollaway nest boxes are used, eggs left

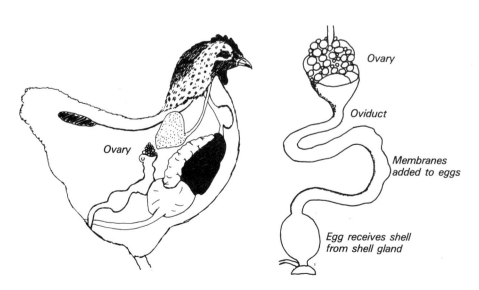

Figure 9.1   The egg laying system.

Collecting the eggs in the central passageway of a large house. Note the eggs in the rollaway nest boxes on the right. Once collected, the eggs are placed in Keyes trays and stored in cool conditions in an egg room. The numbers are recorded daily.

collecting and handling them. On a commercial scale, there is a requirement to do so!

An egg trolley is frequently used by large producers. This can easily negotiate a central collecting passage in a large house and is a convenient way of transporting them to the egg storage room.

Household eggs should be stored until use in a cool pantry or the egg storage area of a domestic refrigerator. Any cracked or soiled eggs should be used immediately and be well cooked.

Eggs that are to be sold should be stored in an adequately insulated and ventilated egg room where the temperature is around 10 to 12°C. The egg marketing regulations state that the temperature must not exceed 18°C. If necessary, a cooling unit or a portable fan can be installed.

Selling eggs

There are three ways of selling eggs, depending on the scale, and it is appropriate to examine these.

Occasional sales of surplus eggs

Anyone can sell their surplus eggs without the need to register with any official body, as long as the following conditions are met:

- The eggs are the producer's own. They are sold only at the following locations: at the producer's farm gate or farm shop, door-to-door sales or market stall.
- They are not graded and sold by graded sizes.

Eggs sold occasionally can be called 'free-range', although this is a trade description, because the authorities have recognised that small, domestic producers are not likely to be keeping their hens in any other way.

too long may be trampled upon and broken, and they will probably become soiled with droppings. In addition, the sight of the eggs may trigger off an unwanted period of broodiness in a bird, or even egg eating. In hot weather, it is obviously not a good idea to leave eggs where they will become warm because of the danger of infection.

Traditionally, a wicker basket was used to carry the eggs, although any clean container that allows them to be transported without knocking against each other is satisfactory. Although the eggs have a protective bloom, it still makes sense to wash your hands before

Bear in mind that the eggs should be fresh and not washed to get rid of soiling. Any with hairline cracks should also not be offered for sale. Eggs are a food product and some local authorities may require occasional sellers to inform the Environmental Health department that they are selling eggs. Where this happens, it is usually a fairly relaxed procedure and there is no inspection.

Selling through a packer

A packer is a distributor who is recognised by REMI as an egg packing station. The distribution company will arrange to collect the eggs of producers who use their services, and then grade and package them on their behalf. They are then distributed to the shops, including supermarket chains.

There are several advantages to using a packer. The producer can concentrate on production, without having to worry about selling produce. The packer will take all the eggs, regardless of size, and sell them by the most appropriate means. Collection is on a regular basis and the packer provides the transport. As he is also responsible for all the grading and packaging, there are no capital costs involved in providing your own grading and packing operations. The disadvantage is that the producer has no control over the prices that he receives for his eggs.

Only those who are operating large free-range flocks would normally be considered by packers, for output must be regular without seasonal fluctuations. This means managing flocks of different ages at different stages of production. A contract with a packer, particularly one who is distributing free-range eggs with a designation such as Freedom Food or Organic, will also require that the producer meets the exacting standards of welfare and general management. Regular checks are carried out to ensure that the standards are being met.

Producers who are supplying free-range eggs to packers must also register with REMI and keep appropriate records. Details of keeping records are given in Appendix 1 of the Reference Section.

Registering as a packing station

Alternatively, the producer can apply for registration and set up his own packing station where eggs are graded into sizes and packed in containers with the description free-range. It is necessary to comply with The Eggs (Marketing Standards) Regulations, 1985 and the Eggs (Marketing Standards) (Amendment) Regulations 1987 and 1991. These ensure that egg quality is maintained for weight, marking, packaging, storage and presentation for sale.

There is no minimum throughput qualification for registration, and the small poultry keeper operating from home can just as easily become a registered packing station as the large unit, as long as the legal requirements are met. This may mean that a room or outbuilding will need to be adapted to meet the requirements. Appropriate items of equipment are available in a wide range of sizes, catering for the small scale producer as well as the large. To qualify for registration, premises must meet the following conditions:

- They must be sufficiently large to accommodate the volume of work being done.
- There must be adequate ventilation and lighting.
- The premises must be easily cleaned and disinfected.
- The area must be suitably insulated to keep the eggs from being affected by excessive heat and cold.
- The area must be used only for handling eggs or for storage of other products which will not affect the eggs.

The premises must also have the following equipment:

- Suitable candling equipment which can be of any recognised type, including hand candling facilities. A hand candler is often used by the smaller producer. This is a device that emits a bright light so that the egg contents can be viewed against it. There is one shown on page 125, although there it is being used to check a fertile egg during the process of incubation.
- A machine for grading the eggs by weight. These are normally for large numbers of eggs and are expensive. They often include built-in candlers. Smaller producers can use a set of scales, as referred to below.
- A device for measuring the air cell in the egg. This is a gauge which is held against the egg when a bright light (candler) is shone through it. It enables the height of the air cell, which is a guide to the age of the egg, to be measured. In a newly laid egg, there is no distinguishable air cell. As the egg cools and ages, the inner cell membranes contract and pull away from the outer ones, leaving a space at the wide end of the egg. By the time a newly laid egg has cooled, the air cell will be around 3 mm. The older the egg gets, the larger the air cell becomes. In a grade A egg it should not exceed 6 mm.
- One or more adjusted balances for weighing eggs. Large producers with an egg grading machine are required to have scales for checking its accuracy. The small producer will be using the scales for grading.
- A device for stamping eggs where use is made of 'date of lay' markings.
- The premises and equipment must be maintained in a good state of repair and cleanliness, and must be free from smells which would taint the eggs.

A separate application must be made if the description free-range is to be used. A Regional Egg Marketing Inspector will visit the premises to check that all the requirements are being met. If not, he will advise on what improvements are needed. Once registration has been approved, a registered number will be allocated to the premises.

Smaller producers of free-range eggs are generally better off doing their own packing and selling direct to their own customers. Free-range eggs from a known local source are often more popular than those in supermarkets. The returns are greater, although there will be the capital costs of setting up and equipping registered premises, as outlined above. However, the costs are not great if small, hand-operated equipment is used.

Packaging and other materials will be higher, without the economies of scale, and the owner must be both producer and distributor, with all the extra work and expense that this involves. Such units are normally family-run enterprises, with a farm shop on the premises and deliveries to local shops, hotels, restaurants, and perhaps even to the smaller multiples. It goes without saying that time needs to be spent on marketing research, going out to meet potential customers and arranging contracts.

Anyone using Special Marketing Terms (SMTs) such as Free-Range must keep special records. This is also the case for those who are producing Organic or Freedom Food eggs. Details are given on page 160.

A franchise operation

The smaller producer may find that franchising is a good way of starting. This type of operation offers a complete 'package' of housing and equipment, birds and feed, as well as providing training. Obtaining planning permission for a producer is also something that a franchiser may undertake.

The producer pays for the building that the franchiser erects on his land, as well as appropriate equipment, birds and feed. He has very little control over costs and margins, and all the elements of the operation, as well as management procedures, are specified by the franchiser. However, all the eggs are collected regularly and taken to the packers. Birds at the end of their productive lives are also bought back. It is highly convenient for the producer for he can concentrate on production rather than having to worry about marketing.

As with any contractual operation, details should be studied carefully before entering into a commitment. An accountant, bank manager or solicitor can advise before contracts are signed.

Egg grading

Producers who have registered as a packing station will be involved in the grading and quality assessment of eggs before they are sold. Free-range eggs are fresh eggs, as referred to at the beginning of the chapter. The EU regulations governing egg sales define the quality of eggs to aim for by a system of grades. The top quality is Grade A. In order to meet the requirements of an 'A' designation, eggs must have the following features:

- *Cuticle* – normal, clean and undamaged.
- *Shell* – normal, clean and undamaged.
- *Air cell* – stationary and with a height not exceeding 6 mm.
- *Albumen* – clear, limpid, of a gelatine-like consistency, free of all foreign bodies of any kind.
- *Yolk* – visible on candling as a shadow only, without clearly discernible outline, not moving appreciably away from the centre of the egg on rotation, free of all foreign bodies of any kind.
- *Smell* – free of foreign smell.
- *Wet or dry cleaning* – not permitted.

There are other grades, but these are not applicable to free-range eggs. Grades B and C, for example, include dirty, damaged or treated eggs.

Shell colour

Brown eggs differ from white in that they have a certain amount of surface pigmentation. This is apparent if you pick up a newly laid speckled egg that is still damp. If you rub it with your finger, the brown markings will be rubbed off. Sometimes even the movement of the hen pushing the egg against the nest box liner will make this happen. Examples are the eggs of the Maran, Welsummer and Speckledy. Once the egg has dried, however, the pigmentation is set and even subsequent boiling will not affect it.

The factor for brown shells is an inherited one and has nothing to do with production methods. White eggs can be laid by free-ranging hens, just as dark brown ones can be produced by battery birds. The egg colour is entirely dependent on the genetic make-up of the chickens. It is surprising that there is still a widespread belief in Britain that brown eggs are better than white, a trend reflected in public shopping habits. In other European countries, such as Spain, white eggs are far more common. Most of these come from hybrids based on the Leghorn. In order to meet the demand for

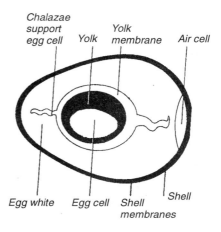

Figure 9.2 Structure of the egg.

brown eggs, breeders of hybrid layers have, in recent decades, made strenuous efforts to develop 'brown' egg strains, based on the Rhode Island Red, that lay comparable numbers of eggs to the more prolific 'white' strains.

Egg size

European Union grading sizes are as follows:

Very Large	73 g and over
Large	63 g up to 73 g
Medium	53 g up to 63 g
Small	Under 53 g

In the USA, they are as follows:

Jumbo	2.5 oz
Extra large	2.25 oz
Large	2 oz
Medium	1.75 oz
Small	1.5 oz
Peewee	1.25 oz

(Please note that 1oz is equivalent to 28g)

Most breeders of egg laying hybrids will provide an egg weight profile for their birds, indicating the percentage of large eggs which can realistically be expected during the laying period. An example of such a profile is shown in Table 9.1.

Feeding obviously plays an important role in egg size and an adequate feed ration must be made available. The cost of compound feeds should also be set against the projected sales of particular egg grades. For example, it may be worth feeding a top quality ration with adequate methione and linoleic acid to produce the maximum number of large eggs, but if a producer has been asked for a regular consignment of smaller eggs it would be sensible to tailor the feed according. Larger

Table 9.1   Egg weight profile and results from a 5,000 free-range flock of Shaver 579

Number of birds	5,000
HHA to 74 weeks	300.22 eggs
Average egg weight	66.2 g
Egg mass	19.87 kg
Feed (Bibby)	117 g
Feed conversion ratio	2.43:1
Mortality to 74 weeks	4.68%
Rearer	Maple Leaf Chicks

Age in weeks	Egg weight
22	50.5 g
26	57.8 g
30	61.0 g
34	63.0 g
40	63.6 g
50	67.0 g
60	68.6 g
70	71.0 g

Source: Shaver 579 Information Bulletin, 1996.

eggs are also more likely to have shell quality problems and the layers can suffer.

It is one of the most difficult balances to get right. In large enterprises, feeding ratios are frequently worked out by computer in order to maintain precise control over a given period. Phased feeds are also available which are tailored to the requirements of the pre-laying, peak laying, mid-laying and end of laying periods. Bear in mind the comments that were made in Chapter 7 about the problems associated with 'rich' feeds, over-production of large eggs and stress leading to unseasonal moulting.

Egg shape

Everyone knows the ideal shape for a chicken's egg. Those which do not conform because they are ridged or otherwise misshapen are likely to be viewed with suspicion. There may also be problems fitting them into egg boxes.

The first eggs from early layers may be small and misshapen, or even large and misshapen, but this is normally temporary. Misshapen eggs may also be an indication of disease, and should be investigated.

Shell strength and texture

Shell strength is important if a high incidence of cracked eggs is to be avoided. Genetic factors are involved to a certain extent, with some stock having a greater potential for producing stronger shells. It is worth checking the results of the European Laying Trials in this respect. The poultry press normally publishes the results each year.

Shell strength and texture both deteriorate as the bird gets older. Feeding also plays a vital role, with an important emphasis on the relative balance of calcium and phosphorus. Research indicates the need for higher levels of calcium and lower levels of phosphorus than those advocated in the past. Concentrate feeds are formulated with these recommendations in mind, and it is not normally necessary to give extra supplies of these minerals. Bear in mind the information given in the chapter on feeding about how calcium may become unavailable because it sifts to the bottom of the feed. Supplying coarse oystershell is an excellent way of ensuring that sufficient calcium is being taken in.

An excess of phosphorus can cause ridging and distortion of the shell.

Yolk colour

Free-ranging birds will have yellow yolks if they have access to good pasture with a range of grasses and clovers. In winter and early spring, when the pasture is not growing, the yolks will be paler. Those of battery caged birds would be pale all year round unless artificial yolk enhancers were used. Consumers wrongly assume that pale yolked eggs must be from batteries while strong coloured yolks are from free-range units. The opposite is often the truth. Organic producers who do not use yolk colour

Ridges on shell

Uneven shell thickness with, possibly, complete lack of shell

Figure 9.3   Variations in shell quality.

enhancers often make a point of including this information on the egg boxes.

Grass and maize meal both contribute to a good yolk colour, but most concentrate feeds also include artificial colour enhancers. In recent years there has been a reaction against the use of artificial colourants, and feeds without these are available.

The Roche scale is the standard way of determining degrees of yellow-orange in the yolk, by comparing it with the scale of colours.

Internal quality

Registered packers have to instigate a 'quality control' programme and examine a representative batch of eggs every so often. This will include candling, or examining some eggs against a bright light in a dark room, and breaking some eggs open to check their appearance and consistency, as detailed in the requirements above. The albumen height, for example, is measured in Haugh units.

Some eggs should also be broken to maintain a visual check for factors such as blood spots or other contamination. If these are discovered, a different management procedure may be called for.

Egg purity

Eggs have a natural and protective bloom or

Eggs of the Speckledy, a hybrid based on the Maran, packaged with an illustration and description of the bird.

cuticle, which helps to protect them from external contamination. Once they are washed, this protection is destroyed, and their purity is compromised. It has already been mentioned that free-range eggs which are offered for sale should not be washed; if they are dirty, they are barred from being sold anyway. Every effort should be made to produce clean eggs by maintaining a high standard of production, collection and storage.

Packaging and labelling

If eggs are being collected by a distributor, they will be packaged at their premises. If the producer is selling them himself and is registered as a packer, he will need to have packaging and labelling equipment and supplies.

Egg cartons are widely available for free-range producers who are selling their own eggs. They are available plain or pre-printed. There is also the facility for having them printed with the producer's own details, or on a smaller scale, labels can be added for this purpose.

Free-range eggs may be sold in small cartons that are marked 'Free-Range Eggs'. They may also be described as Very Large, Large, Medium or Small, depending upon their weight. A description on the box might read:

> # FRESH
> # FREE-RANGE
> ## Large
> Six Eggs
> Class A

The producer's name and packing station number are included, together with any graphic illustration or advice to the consumer, such as when the eggs 'Best before'. Registered Organic or Freedom Food producers will also be able to use the appropriate logos and descriptions. Anyone wanting to sell eggs, whether graded or not, should obtain full details from the Regional Egg Marketing Inspector. Legislation can and does change.

Preserving eggs

The following information is geared to the domestic flock owner who may be interested in traditional and modern methods of food preserving. The only methods that I would recommend are the freezing or pickling of absolutely fresh, clean and undamaged eggs. The other methods are given for information only. Experimenters do so at their own risk.

Waterglass

Waterglass is a mixture of 1 part sodium silicate to 9 parts water. Some chemists still sell it. Place it in a high-density, food-quality, plastic bucket and cover with a lid. The fresh eggs are then carefully placed in the solution every day until the bucket is full. No part of the eggs should be above the surface. They should keep for up to six months in this way, although some may go off if they had hairline cracks.

Frozen eggs

Eggs can be frozen for about three months, but it is necessary to remove the shells or they will burst. The most convenient way is to break the eggs into an ice-cube tray, keeping the yolks separate from the whites. Add a little salt to those yolks which will be used in savoury recipes and sugar to those which are to go in cakes. Both salt and sugar will stop the yolks becoming sticky when

they are defrosted. Remove the frozen squares from the ice-cube trays, package and label them appropriately. Two white squares and one yellow will equal one complete egg.

Alternatively, surplus eggs can be used to bake sponge cakes or sponge bases, which can then be frozen until required.

Pickled eggs

Eggs can be pickled in cider vinegar that has previously been boiled for a few minutes and then cooled. The eggs should be hard-boiled and then shelled before being placed in the cooled vinegar. They will keep in a closed glass jar for about two months. Cider vinegar is used in preference to malt vinegar because it is milder. Spiced vinegar can be made by putting pickling spices in a muslin bag in the vinegar while it is heating. The bag is then removed when the vinegar has cooled, and before it is used.

Other methods of preserving

The following methods have been suggested by readers of *Home Farm* magazine:

- Tie fresh eggs in a muslin bag. Plunge bag into boiling water for 6 seconds only. Draw out at once and leave to cool. Then store in a very dry place. They will remain fresh for 3 months. In a refrigerator they remain fresh for 6 months.
- Dry cheap cooking salt in the oven and then pack eggs in a crock with salt above and below. They will keep for 6 months.
- Rub clean, fresh eggs with lard or paraffin wax.
- Paint clean, fresh eggs with thin gum arabic (equal parts of gum arabic and water).
- Place eggs in their shells in brine. They will keep for up to a month.
- Store eggs in their shells in lime water of 2 parts slaked lime and 1 part salt to 16 parts water.

CHAPTER 10 # Table Birds

Foods of the future will have to convey a theme of fitness and fun, be convenient and of high quality.

Colin Groom, ADAS Marketing Adviser, 1986

Traditionally, it was heavy breeds such as Light Sussex, Plymouth Rock, Indian Game (Cornish) and Dorking that were used for the table. Then, the Cobb, a white-feathered broiler, was bred from a white dominant strain of Plymouth Rock crossed with Indian Game (Cornish). Although bred for the intensive sector, it can adapt to outside conditions where the exercise and less intensive feeding regime slows

down its growth. In this way, there are less likely to be leg problems. I often raised Cobb 500 females on free-range but gave them a lower protein ration of around 14%, together with a grain ration.

In recent years, colour-feathered varieties of breeds specifically reared for free-range have become available. In France, they have developed primary breeders that can be used to produce a wide range of coloured feathered birds to suit different markets. (Areas of France have traditionally had their own local type and colour of table bird). The principle is that small, prolific hybrid hens are crossed with a pure-bred terminal sire. The females have a recessive gene so that the progeny always resemble the father, but have the plumpness, short-leggedness and productivity of the mother. Examples of primary breeders include Sasso SA51, Sasso SA31, Hubbard-ISA Redbro and JA57. Sasso have also developed a Transylvanian Naked Neck primary breeder because it has a useful gene for reduced abdominal fat. Some of the colour-feather table breeds based on the primary breeders, and available as day-olds, are Poulet Gaulois, Farm Ranger and Poulet Bronze.

There are also white-feathered birds that have been developed for free-range. They include Hybro and Sherwood White.

Redbro, a slow-growing strain suitable for free-range table bird production.

A commercial flock of Sherwood Whites in ideal forested conditions such as the domestic chicken's wild ancestors, the Jungle Fowl, would have enjoyed. (*Premier Poultry*)

Managing

Any of the houses that are used for laying chickens (apart from the very large static houses) will be suitable for rearing table birds. Before they go in there, however, they need brooding conditions.

Most table birds are bought as day-olds. This is the cheapest way of acquiring them, but if they are bought 'as hatched' (AH) a certain proportion will be males. Buying them sexed is more expensive, and the sexing will not be 100 per cent accurate. An alternative is to buy them at around six weeks old when they are hardy, and then raise them to killing weight at around twelve weeks.

Those who are registered for producing organic table birds can buy them in as day-olds. These may either be produced on-site or bought in from an organic hatchery.

Brooding conditions

Day-old chicks will need protected, warm conditions until they are full-feathered and hardy. Purpose-made brooders are available, but these are not essential. What is more important is the provision of a dry, rat-proof house with good insulation and ventilation. The floor should be covered with a layer of wood shavings or similar warm litter. Initially, a small area of the floor should be made available to them, with temporary walls of blocks, straw bales or even cardboard, to protect them from draughts and to keep them confined to the heated area. These walls can gradually be disposed of as the chicks get bigger.

A brooding lamp is needed to provide the chicks with warmth. Hang it above the confined area, at a height that will be comfortable for the chicks. If it is too low it will be too hot, and the chicks will move to the outer edges to get away from the heat. If the lamp is too high the chicks will be cold and will huddle in the middle. Brooding lamps are available powered by electricity or propane gas. Dull emitter lamps will provide heat without stressful glare.

Housing

As the chicks grow, their need for warmth from the lamp gradually diminishes. It can be raised higher and higher, until it is no longer required.

Once they are off heat, the young birds can be placed in a house similar to that of free-range layers, as referred to earlier. They need to be protected against foxes, either with a tall fence or with electric fencing. If a large number of day-olds are reared, they can be split into different housing groups, to comply with organic standards.

My own system of housing table birds is shown in Figure 10.1. It includes an ordinary house of the kind that can be moved from one area to the next without too much difficulty. A chicken wire fence was erected in front of the house to make a straw yard and the birds were confined to this area if weather conditions were poor. In fine weather the gate was left open so that they could range on grass in the rest of the field but, as I did not have electric fencing for a number of years, considerable care had to be

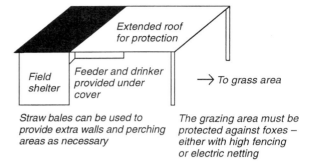

Figure 10.2　An alternative method of housing table birds.

exercised in relation to foxes. One year, the fox came before dark and I lost nearly a third of the batch.

Feeding

The chicks will need feeders and drinkers, either suspended or placed on the ground.

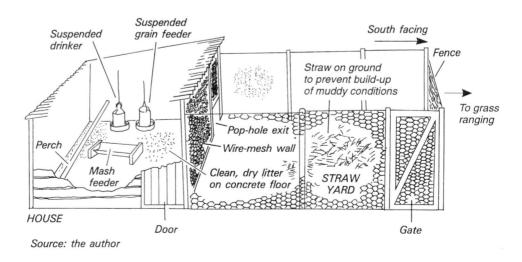

Figure 10.1　Example of a small free-range table bird management system.

One of the new range of moveable houses for commercial layers or table birds. Trees provide shelter and a sense of security, while electric netting controls access to pasture and keeps out the fox. (*Associated Poultry Services*).

Chick crumbs are given in the initial stages, a compound feed small enough for ease of picking. Most chick crumbs contain a coccidiostat, but it is possible to buy ones without this. Organic standards allow the use of Amprolium if coccidiosis should occur. All other in-feed medications are banned.

There are companies that specialise in the provision of natural and organic feeds. These are obviously more expensive than ordinary feeds, but are recommended for anyone who is rearing table birds domestically or commercially. They are essential for commercial organic production. Unsprayed pasture is also a requirement of the organic standards.

Those selling table birds with specfiic descriptions such as Free-range, Organic or Freedom Food will need to ensure that proper conditions are met at all times.

The feed ration of table birds can be made available on an ad-lib basis either inside or outside the house. When I visited a poultry farm in France I was interested to see that their food was put in large outside hoppers. These were placed all over the site to encourage wide ranging, and the birds were fed exclusively on maize grown on the farm. This was chopped up using an electric grain grinder equipped with kibbling plates so that the resulting pieces were of an appropriate size for the birds. Table birds which are fed on such a ration are referred to as 'corn-fed', and they develop golden skins as a result.

Water and insoluble grit should be made available to the birds at all times, with shading being provided for outside feeders and drinkers in hot weather.

Compound organic rations for table birds contain wheat, fish meal, full fat soya, field beans, lucerne and a mineral and vitamin supplement. This is in addition to a whole-grain scratch feed of wheat. Broiler rations for the intensive sector also contain antibiotics to promote growth and to keep the birds from succumbing to infection. Avoid them at all costs!

Killing

Naturally reared table birds will be ready for slaughter at 12 weeks. By this time they are around 1.8 to 2.3 kg (4 to 5 lb). Intensively raised broilers will achieve this weight in half the time.

Commercial rearers of organic poultry can either slaughter and process the birds on the farm or send them to licensed slaughterers. The latter is obviously more expensive, but planning permission may be necessary if it is done on the farm. Planners often regard a 'food processing' activity as being separate from farming. (See the planning section in Chapter 5.)

If killing, plucking and dressing are done on site, it is essential that proper procedure has been followed. ATB-Landbase has a database of training organisations that are registered with them. Essential reading is *The Code of Practice for On-Farm Slaughter and Marketing of Poultry* available from the Ministry of Agriculture. Equally essential is *Practical Slaughter of Poultry: A Guide for the Small Producer* from the Humane Slaughter Association.

The following legislation will also need to be adhered to: Welfare of Animals (Slaughter or Killing) Regulations 1995; The Poultry Meat, Farmed Game Bird Meat and Rabbit Meat (Hygiene and Inspection) 1995; The Food Safety Act 1990.

Birds can be slaughtered by neck dislocation in emergencies or if the meat is for home consumption. If it is to be sold, the birds should be killed by stunning, followed by neck dislocation or decapitation. A humane stunner can only be operated by a trained and licensed person.

If the birds are to be slaughtered on site, they should be housed in protected conditions and have access to fresh water at all times. Grain should not be given for 24 hours, but dry mash can be given until 6 hours before slaughter. Such a practice will ensure that the gut is relatively clear, without causing stress to the birds.

Plucking

There are three methods of plucking: traditional hand plucking, machine plucking and wax plucking. The method used will depend on the scale. Small producers will tend to use the traditional method, although small machine pluckers are on the market. Part-time experienced help is often available in rural areas.

Wear an overall and head covering, and work in light, airy conditions. A barn with a sealed concrete floor is ideal. The feathers will come out more easily while the carcase is

Feather plucker, suitable for table fowl.
(Stock Nutrition)

still warm. The skill lies in removing the feathers without tearing the skin. The sequence will be a matter of personal preference. Some people start with the breast feathers, then go on to the back, sides and finally the wings. Once the birds are plucked, they can be hung in a cool room for 1 to 2 days.

Processing poultry

The Environmental Health department of the local authority needs to be informed if food processing is taking place. They will also offer advice to those who are setting up a farm enterprise which involves the processing of food. It is important to be familiar with all the food regulations that apply, and Environmental Health are the best people to advise.

Hygienic conditions are essential at all times and obviously every effort should be made to avoid cross-contamination during the processing of poultry. Clean overalls and head cover should be worn and there should be access to hand-washing facilities at all times. The flow chart shown in Figure 10.3 indicates a system where the initial plucking

and gutting operations are kept separate from the subsequent dressing of the birds. Feathers and viscera should be disposed of immediately, either by incineration or by removal by a contract firm if the scale warrants it.

To eviscerate a bird, a good, sturdy beech table is ideal, although any surface that can be scrubbed after use is suitable. Have plenty of muslin cloths for swabbing soiled areas, and a number of buckets or large collecting utensils for the entrails.

Cut off the head and make a cut in the skin along the back of the neck. Use poultry secateurs to cut through the neck and remove it. A piece of muslin placed around the neck will make it less slippery to get hold of. Enlarge the incision in order to get the hand in after the neck has been removed, and take out the gullet. Then fold over the flap of skin.

Turn the bird around and make a circular incision around the vent so that it can be pulled clear. Take care not to pierce the rectum. As the vent is pulled clear, the intestines follow after and can be dropped into a bucket. Enlarge the opening enough to get your hand in and draw out the gizzard, liver, crop and lungs. The liver, heart, neck and gizzard can be retained, for

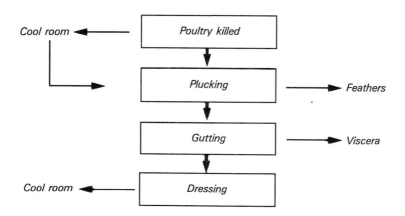

Figure 10.3   Flow chart for a table bird production system.

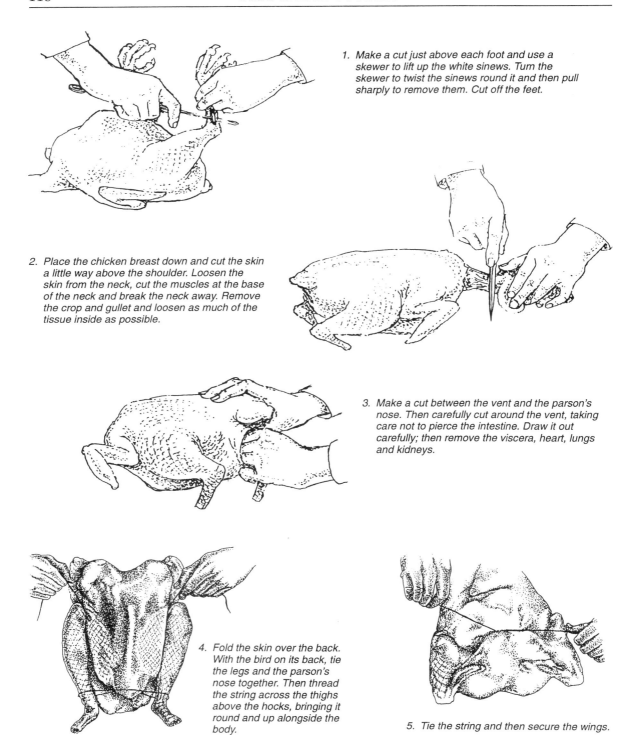

1. Make a cut just above each foot and use a skewer to lift up the white sinews. Turn the skewer to twist the sinews round it and then pull sharply to remove them. Cut off the feet.

2. Place the chicken breast down and cut the skin a little way above the shoulder. Loosen the skin from the neck, cut the muscles at the base of the neck and break the neck away. Remove the crop and gullet and loosen as much of the tissue inside as possible.

3. Make a cut between the vent and the parson's nose. Then carefully cut around the vent, taking care not to pierce the intestine. Draw it out carefully; then remove the viscera, heart, lungs and kidneys.

4. Fold the skin over the back. With the bird on its back, tie the legs and the parson's nose together. Then thread the string across the thighs above the hocks, bringing it round and up alongside the body.

5. Tie the string and then secure the wings.

Figure 10.4　Dressing a bird.

many customers like to have these for making stock or pâté. The gizzard will need to be cut open and the stones and tough membrane discarded. The giblets can then be put in a plastic bag ready for placing inside the carcass when everything else is finished.

Next make an incision just above each foot but do not cut right through. Break the leg bones by snapping them over the edge of a table and the leg tendons will be revealed. They look like white elastic bands and need to be removed because they are very tough.

The easiest way of doing this is to have a purpose-made tendon remover screwed into the wall. This is rather like a double hook and when the foot is placed in between the prongs, and the carcass drawn down sharply, the tendons are pulled out. Tendon removers are available from poultry equipment and farm suppliers.

The bird should be stored, breast side up, in a cool room ready for sale or use. Plastic bags, ties and labels are widely available for presenting the oven-ready chicken to its best advantage.

CHAPTER 11 # Breeding and Rearing

At the present time, Imogene is secluded in a dark barrel, lavishing her affection upon two pieces of brick, an old axe head and three discarded china door knobs. I admire her indomitable perseverance, but I don't believe anything will ever come of it.

Editor of Rockland Tribune, *Maine, USA, 1895*

Breeding chickens is one of those activities that is either on a very big or a very small scale. The breeding and rearing of hybrid birds is the province of large breeding companies, while that of pure breeds tends to be in the small sector. The coverage of breeding in this chapter is therefore applicable to those who are interested in pure breeds, particularly in the improvement of the old utility breeds. The section on rearing is relevant to anyone who rears young poultry, whether it be day-old chicks as replacement pullets, pullets for sale or young table stock.

Anyone with a breeding flock of 250 birds or more must arrange to test them regularly for salmonella, under the requirements of The Poultry Breeding Flocks and Hatcheries Order 1993.

Breeding stock

The most difficult problem facing those who wish to try to breed replacement birds is finding a good male. Most of the large fowl in the pure breed sector have a poor level of fertility, their productive qualities having taken second place to show characteristics over the years. Few commercial, utility strains now exist.

Within the hybrid layer sector, there are superb breeding birds but they are not usually available for sale. It is only their progeny which are sold, usually as female day-olds. If a good utility male does become available, he should be treated like gold dust! Hybrids will not necessarily breed true, but that is no reason why they should not be crossed. A good male, for example, can be used to introduce hybrid vigour into a pure strain that has lost its productivity. Careful selection of the progeny over several generations can bring back a breed with renewed vigour. An example of this, but in reverse, is how breeders of the Maran are using some of the new Speckledy hens to cross with their pure breed males to try to improve productivity.

If it is possible to obtain hatching eggs of hybrid laying stock, a certain proportion of the chicks will be males. These can be raised, with the best being kept as breeders. The males from a brown hybrid breed are light while the females are brown. When they are mated in the second year, the sex-linked cross will be reversed. Unfortunately, commercial semen for artificial insemination, although used in the commercial poultry sector, is not available to the small poultry keeper.

Any bird that is to be used for breeding should have a blood test in order to ensure freedom from all diseases which can be transmitted to the chick via the egg. A vet will take blood samples from birds concerned and arrange for them to be tested. Once tested and found to be clear, the breeding stock should be kept quite separate from other poultry. It is first necessary, however, to identify which pullets are suitable for keeping as breeders.

As far as selecting for egg production is concerned, it will be necessary to record the number of eggs produced by particular birds in order to choose the best layers. Where only a small number are concerned, it may be possible to identify individual eggs as belonging to specific birds. It is surprising how often this is possible. The best way of doing it, though, is to use a trap-nesting system.

Trap-nesting

A trap-nest is one that allows a hen to go into a nest box, but is so designed as to prevent her getting out again. She cannot escape until you come to release her, so it is possible to see which hen has laid which egg. The birds are identified by numbered leg rings. As the rings are also made in different colours, the various lines of birds can be identified. Leg rings are available from poultry suppliers or from the Poultry Club of Great Britain, which has a ringing scheme. It is important to check trap-nests frequently so that the hens do not spend more time in them than absolutely necessary. Trap-nests are not difficult to construct, as Figure 11.1 shows.

Record the results every day! How this is done is not important as long as the information indicates the number and quality of eggs for each hen. These factors may include shell shape and strength, colour and markings. It may simply be a matter of entering the details on a calendar or diary. A notebook can be ruled off into columns to enter the details, or a computer can be used to store the data. See page 156 for further details on keeping records.

The difficulties of selective breeding should not be underestimated. It is a long process requiring a great deal of time and patience. It has been said that at least 10 generations must be bred in order to achieve a substantial result. Also necessary is a strong resolution not to breed from inferior birds.

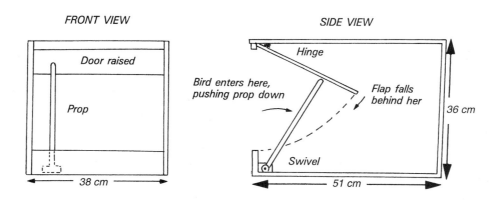

Source: Katie Thear. Incubation: A Guide to Hatching and Rearing. Broad Leys Publishing.

Figure 11.1   Trap-nesting.

Line breeding

Line breeding is the establishment of several different, unrelated lines of birds, with a different cock for each line. If each line is given a different colour for identification and recording purposes, it will be a simple matter to tell them apart. For example, one line could be given red leg rings, while another is given green ones. As each ring has a different number, it is easy to establish the identity of each bird within the line. When chicks are hatched from a particular mating, a dab of indelible ink of the appropriate colour can be painted on the top of the head. Later, they can be given leg rings.

Occasional in-breeding can take place if it is particularly desirable. For example, a particularly good hen could be mated with one of her sons, if he is also outstanding, but the females of that cross should be mated with an unrelated cock if breeding is to take place again. The original hen should not be mated with any males from that cross, if this can be avoided.

Housing and management of breeders

A breeding cock should have his own house and run to which the females are introduced as necessary. Nine or ten hens of light breeds can be put in at the same time, but six or seven is the maximum for heavy breeds. The females should all be in-lay when they are introduced. Hens of heavy-feathered breeds should have some of the feathers around the vent clipped back, so that there are no obstructions.

If the spurs of the cock are long, it is advisable to trim them back a little with clippers and then keep them filed down. Do it carefully, making sure that only the tip is blunted or it will bleed. Spurs are the main fighting weapon of the male and their length was much prized in the illegal days of cock-fighting. There is no truth in the old belief that spur-trimming leads to infertility. If spurs are left long, the hens may suffer

considerable damage, including ripped sides. Cocks can be dangerous, particularly to children, and should always be kept confined in the house and run.

Breeding birds must be fed well, for any deficiencies will show up in the chicks. A lack of vitamin B_2, for example, can produce chicks whose down feathers have clubbed ends. In extreme cases, they may also have curly toe paralysis. Make sure that the birds have a good ration of compound feed, grain, water, grit and crushed oystershell. Breeder's rations with the appropriate minerals and vitamins are now available in relatively small quantities. Vitamin and mineral supplements can be added to other rations if one specifically for breeders is not available. On a small scale it is also possible to improvise. For example, a ground-up yeast tablet or a little Marmite is a good source of vitamin B_2. Figure 11.2 lists some other possibilities.

If the breeding house is equipped with nest boxes the hens will lay fertile eggs there. They can then be collected and stored ready for incubation. Keep them in a cool place (12° to 15°C) in an egg box, with the blunt ends of the eggs uppermost. Keep the box at a slight tilt, and reverse the tilt every so often. Ideally the eggs should be incubated as soon as possible, but certainly no later than a week, although a 14 day interval can still produce some viable eggs.

Incubation

Incubation is the process of development of the fertilised egg into a fully grown chick ready to hatch. It takes twenty-one days for this miraculous transformation to take place, and the stages are as follows:

4th day: leg and wing buds form
6th day: subdivision of legs and wings
8th day: formation of feather tracts
15th day: calcification of bones complete

Problem	Cause
Eggs do not develop Eggs clear when candled Eggs gone bad	Infertile eggs or left too long before incubating. Cracked or damaged eggs. Possibly diseased. Disinfect incubator before use and dip eggs in sanitant. Ensure that parent stock is fertile.
Partial incubation with a 'blood ring' in the shell	Temperature fluctuations – Check the thermostat. Possibly a bacterial infection – Attend to hygiene. Possibly a viral infection – Ensure that parents are healthy and free of inheritable diseases by having them blood-tested.
Chicks hatching early, often with blood on their navels	Temperature too high – Check thermostat.
Weakly chicks	Temperature too high – Check thermostat. Poor or inbred breeding stock. Deficiency of essential amino-acids – Feed parents adequately.
Deformed chicks	Too much inbreeding. Hereditary diseases.
Beak or splayed leg deformities	Not enough calcium, phosphorus or vitamin D in parents' diet – Feed adequately. Also give oystershell grit and small amounts of cod liver oil.*
Curly toe paralysis	Vitamin B_2 deficiency in parents – Allow grass ranging or feed supplement of chopped boiled egg, Marmite, yeast or ground-up yeast tablets.*
Clubbed ends to down feathers	Vitamin B_2 deficiency in parents – Feed them supplement as indicated above.
Inability to coordinate movement (Crazy chick disease)	Vitamin E deficiency in parents. Too much maize given at the expense of wheat – Balance diet and give parents a supplement of wheat germ extract.*
Developed chicks 'dead in shell', sometimes with unabsorbed yolk in abdomen	Too high a level of humidity or of temperature at critical hatching time. Deformities of beak may prevent chick from pecking its way out of shell. This is the result of hereditary disease, inadequate parental feeding or in-breeding.
Generally low hatching rate	Any of the above reasons. Possible lack of vitamin B_{12} or shortage of trace elements in parent's diet. Feed properly and give supplement of ground-up multi-vitamin/mineral tablets.*

* A mineral/vitamin supplement will cater for any deficiencies.

Figure 11.2 Incubation problems.

16th day: beak, claws and leg scales formed
19th day: chick complete and begins to draw on yolk sac reserves
21st day: hatching

There are two choices when it comes to incubation: relying on a broody hen or using an incubator. The first option is obviously only applicable to the small scale, and it is a popular method with small poultry breeders. The pleasure of seeing a hen with her newly hatched brood is appreciated by many.

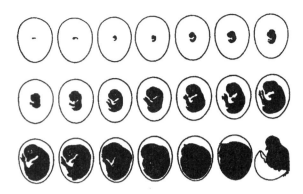

A small table-top incubator with automatic egg turning and thermostatic controls. *(Brinsea)*

Figure 11.3   Stages in incubation from day 1 to day 21.

Incubators

There is a wide range of incubators available, from the small table-top model to the large free-standing one. For small numbers, a table-top model is sufficient, while larger numbers would be more appropriate in a cabinet incubator. This has the door opening at the front, and has several shelves. Some models have a separate hatching compartment because the temperature and humidity conditions required for incubation and hatching are slightly different.

There are two kinds of incubator: *still-air*, where air circulates by normal radiation and convection through the air, and *forced-air*, where a fan assists in the air circulation. The latter is normally found in a large incubator.

Most incubators have egg turning facilities, either automatic or hand-operated. The latter is a feature that allows all the eggs to be turned at the same time. During the course of incubation, eggs need to be turned regularly to prevent the embryo sticking to one side of the egg. With older incubators it is necessary to turn each egg by hand, and mark them with a cross, so that they rest on alternate sides.

The optimum temperature for incubation is 37.7°C at the centre of the egg. For hatching

A small, well-equipped cabinet incubator with automatic controls. *(Curfew)*

Humidity controller that can be used in conjunction with many incubators. Thanks to the greenhouse effect, humidity levels in the air are higher than they used to be, making incubation more difficult. *(Curfew)*

it is a degree less. Electronic thermostats are now available in most incubators, and they are much more sensitive to temperature changes than the old wafer type of thermostat. Humidity should be kept to 52 per cent during the incubation period and increased to 75 per cent for hatching. It is much easier to control humidity if the incubator is placed indoors at room temperature, rather than in a cold, damp outhouse. My hatching percentages were transformed when I placed my incubator in a spare room indoors.

The best way of keeping an eye on humidity is to candle or weigh the eggs during the incubation period. With a candler, a bright light can be shone through the egg enabling the viewer, not only to see if the egg is fertile and developing, but also if it has the right amount of humidity. The size of the air cell in the egg is linked to the amount of water. Figure 11.5 shows the relative size of the air cell at different stages of incubation. If it is too small, the humidity is too high. If too big, the humidity level is too low. Candling also enables infertile eggs to be identified and removed from the incubator.

The key factors with an incubator are to follow the manufacturer's instructions, and to maintain hygienic principles. Place it in a sheltered area where the temperature does not fluctuate too much, and where it will be safe from knocks. Have the incubator up and running for at least 24 hours before using it. Dip the eggs in an egg sanitant before incubating them, to make sure that they are free of pathogens. After the incubator has been used, clean and disinfect it thoroughly before using it again.

Avoid 'helping' the chicks out of the shell. Most will 'pip' the shell and hatch without any difficulty. Sometimes it may appear that the chick is struggling to no avail and there is a great temptation to interfere by helping to break the shell for it. This should be avoided, for it is possible to rupture the blood-vessels in this way. Only when it is quite obvious that there are problems, after all the others

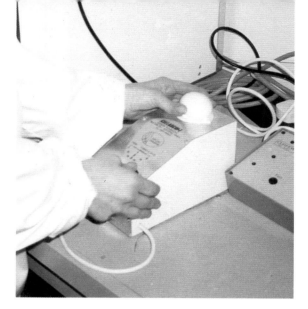

A candler being used to examine the internal contents of the egg.

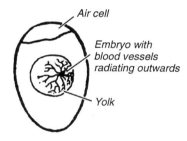

Figure 11.4   Fertile egg candled at 5 days. The embryo can be seen developing as blood vessels radiate across the yolk.

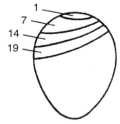

Figure 11.5   A hen's egg showing the size of the air cell at the appropriate number of days into incubation. The best way of checking humidity is to candle fertile eggs and compare the size of the air sac with this standard. Too small – humidity too high.

Chicks hatching in a small incubator.

Brooding conditions. The chicks on the left are newly hatched. The ones on the right are older.

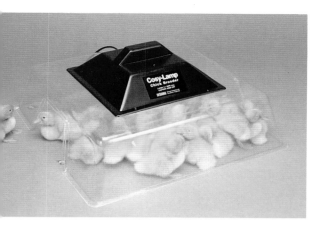

Small brooder for newly hatched chicks or ducklings.
(*Brinsea*)

have hatched, and dried off, should any attempt be made. Sometimes a little warm water bathed around the shell will enable it to break, particularly if the internal membranes have dried and stuck to the chick. When doing this, take great care that water does not go into the chick's nostrils and drown it.

Once hatched, the chicks will soon be on their feet and the down feathers will dry. They should be left for 24 hours before being moved to a brooder. They do not require food at this stage because they have the remnants of the yolk inside them as a food store. The next day they can go into a brooder with a dull emitter lamp to keep them warm, and chick crumbs and water to welcome them into the world.

There will always be some problems during incubation and it should be expected that a certain proportion of eggs will not hatch, and that some chicks will die. The reasons for this are many, and they are impossible to predict with any certainty. Figure 11.2 listed some of the more common problems, but it should be regarded only as a general guide.

The broody hen

Broodiness is the natural tendency of a female bird to sit on, incubate and ultimately hatch a clutch of eggs. It shows itself in the hen's posture and behaviour. She will remain sitting on a nest or in a dark corner, sometimes on the eggs of other birds. She will resist efforts to remove her with loud squawks of protest and a ruffling of feathers to make herself look larger and more aggressive. Her breast area will feel unusually hot to the touch. She will often peck at anyone trying to move her, and feeding time may even lose its appeal. When lifted from the nest and placed on the ground she will either remain there with feathers ruffled and wings slightly outstretched, or will walk around, clucking volubly, looking like a feathery ball with hunched head and neck.

If a hen has become broody but is in an unsuitable place, she will need to be moved to a situation where she will be dry, warm and protected from predators. A small coop and run is ideal. The best time to move her is at night, so that there is less likelihood of disturbance. The eggs she is sitting on can be taken from beneath her, one by one, and placed in the new nest. The hen herself is then gently lifted with both hands firmly around her body so that the wings are confined. On no account allow her to flap her

Croad Langshan cross hen which has become broody in a strawed area adjacent to a building.

The correct way to hold a chicken.

wings in case she cools herself down or panics. Place her on the eggs in the new nest and close the door for the night. In the morning if she is sitting contentedly and shows no sign of restless agitation or wanting to escape, her broodiness cycle has remained unbroken and she has accepted her new quarters.

It is often the case that a broody hen is required to incubate the eggs of another bird. Her own eggs will therefore need to be removed and the new ones substituted. If the hen is also being moved to new quarters, the substitution should be delayed until she has settled down. The danger is that if she is suddenly placed on a clutch of cold eggs she will reject them because she knows that they are not the one she was sitting on. This is not always the case, but it can happen. The best way is to wait until she is sitting contentedly, then to remove an egg from beneath, while slipping a new one under at the same time. Do this gradually until all the eggs have been substituted. If the new eggs are marked in some way, there is no danger of confusing them with the old eggs. The number of eggs that a broody hen is given will depend on her size. A normal hen will accept about a dozen.

The broody needs to leave the nest at least once a day to relieve herself, get exercise and to eat and drink. Usually she will do this for about ten to fifteen minutes, and food and water should be made available for her at the same time each day. Some hens are awkward, in that they refuse to move. When this happens, it is best to lift the bird off the nest and place her near the food and water. She will complain loudly, but will usually have something before going back to the nest. If she refuses the food and goes straight back to her eggs, leave her. Eventually, hunger will force her to eat and if the same pattern is repeated later, she will probably eat and drink then. It is quite common for broody hens to sit tight all day, then to leave the nest for a short while in the late afternoon. If a hen establishes this pattern for herself, it is best to follow her lead and to provide food and water at that time. Give her a quick dusting of lice and mite powder when she has been sitting for a day. This should be done when she has left the nest; simply puff a little powder under her wings around her neck and bottom. It will prevent any problems with external parasites occurring during a time when she will not be taking dust baths.

It is rare for a broody to foul her nest; she will usually relieve herself when she leaves the eggs in order to feed. There is sometimes an awkward one that will foul the nest, increasing the likelihood of disease if droppings adhere to the eggs. Such birds are usually the ones who refuse to move when food is provided for them. Dealing with such birds is quite easy, as there is a trick to make them relieve themselves. Lift the hen from the nest and grasp her by the legs, then throw her up in the air while retaining a firm grip of her legs. She will flap her wings once, as she is placed on the ground, and she will immediately deposit droppings. This technique probably works by stimulating a basic reaction to the threat of danger.

As a general rule, a hen will only become broody if she is ready to do so, but it is sometimes possible to provide conditions which will help to trigger that response. Make a nest in the warmest, darkest corner available and place some 'pot' eggs in it. Make sure that the hen or hens of your choice know that the nest is there and then wait and see if one of them falls for it. If a hen does become broody and adopts the artificial eggs, fertile ones then can be substituted. This method is often successful when the weather is warm, but the problem is that broodies are usually required early in the season when the weather may be cold.

It may not always be convenient for a hen to become broody, particularly if she is supposed to be laying. On page 20 reference was made to the technique for stopping broodiness.

The broody hen that hatches chicks is responsible for brooding them. There could be few better rearers. The protective instinct of the mother hen with her chicks is legendary, and it is delightful to see her spreading her wings over them to protect them from harm.

The best way of housing them at this time is in a small fold unit, which can be moved on to a fresh piece of ground each day. It should have a house at one end into which the hen will retire with the chicks at night, and a protected run so that they are able to have the benefit of the sun during the day. If the fold unit has a wire mesh floor which rests on the grass, they will be protected against rats which might otherwise burrow underneath. A starter ration of chick crumbs should be provided for them, as well as water in a drinker. Gravity-fed drinkers are preferable to open containers, for the water is less likely to become contaminated by droppings.

The hatching of chicks in an incubator may coincide with the availability of a broody hen that has been sitting for a couple of weeks. It may be possible to introduce the chicks to her so that she thinks they are hers and adopts them. The introduction needs to be done carefully. Slip a new chick underneath her in such a way that she does not see the chick beforehand, then at the same time, remove an egg. Slip in a few more chicks, then stand back to see how she responds to them. If she begins to respond by clucking to them and shows no signs of trampling on them or abandoning them, the introduction has been successful. After a couple of hours introduce the rest of the chicks, while removing the remaining eggs and see if all is well. If a hen does accept chicks in this way, it will save you the trouble of having to provide brooding accommodation for them.

The hen will look after her brood for around five weeks and will then gradually lose interest in them. By this time they are fully-feathered and hardy.

If the chicks are from an incubator and have no foster mother, they must be placed in brooding conditions.

Raising day-olds

The details of brooding conditions have already been outlined in Chapter 10. The aims may be different from those of the table bird producer, but the initial rearing management of day-olds does not vary a

A breeding pen of White Silkies. They make excellent broodies.

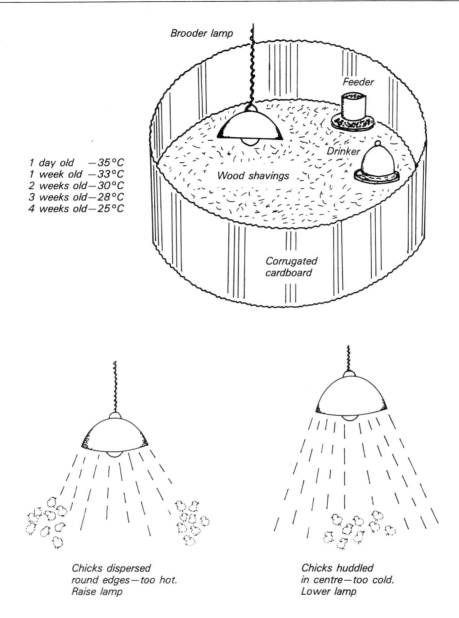

1 day old   −35°C
1 week old −33°C
2 weeks old−30°C
3 weeks old−28°C
4 weeks old−25°C

Chicks dispersed
round edges−too hot.
Raise lamp

Chicks huddled
in centre−too cold.
Lower lamp

Figure 11.6   Rearing day-old chicks.

great deal. Nor does it vary with numbers. The general principles of rearing 5,000 chicks are the same as for half a dozen. They all have the same requirements of shelter, warmth, food and water. It is merely the scale that differs.

One of the many branches of the poultry industry is that of pullet rearing, where day-old chicks are bought from one of the breeding companies and raised until they are anything from 16 to 21 weeks of age. At this stage, depending on the needs of the buyer, they are sold as pullets coming up to lay. From an early stage they should have access to perches and to range conditions so that they are 'house-trained' and have the opportunity of developing natural immunity. Replacement pullet rearers normally come to an arrangement with a breeding company to act as agent for that particular breed.

There may be those who buy day-olds as their own laying replacements, and not for resale to others. This is certainly cheaper than buying point of lay pullets, but it does mean that housing and brooding equipment must be available for them. Not everyone has the buildings and equipment that are required for rearing.

The chicks are usually injected against Marek's disease. Other vaccinations to protect pullets include those for Newcastle disease, infectious bronchitis and endemic tremors. Consult your veterinary surgeon for

A range of pure-breed chicks in protected brooding conditions, although no longer requiring artificial heat.

advice, particularly where local conditions may be especially conducive to some diseases. Gumboro disease used to be a problem that was only found in large broiler houses, but in recent years it has moved into other sectors. It should be pointed out that

Chicks at the hardy stage. They are fed on chick crumbs, gradually changing to a grower's ration before starting a free-range layers' ration when they are at 18 weeks. *(Brettell and Shaw)*

birds for the organic sector are not routinely protected in this way.

Vaccination of poultry is a technique which can be learnt at a poultry management course. It is a technique which must be exercised with care, particularly since the incidence of accidental self-injection is comparatively high.

Salmonella testing is also required where more than 250 breeding birds are involved.

Pullets

Once the pullets are well grown, they will follow different systems of management, depending upon circumstances. Laying pullets will either be sold to egg producers or transferred to the owner's laying house at around 18 weeks of age. From 20 weeks they can be given artificial light to supplement natural light and encourage them to start laying, although this depends upon the time of year. Table birds may be sold at around 6 weeks to those who wish to rear them to full weight, or kept as free-range broilers on the original site.

Pure breeds are often sold as breeding trios of one male and two females at any age from 6 weeks onwards. The male should not be too closely related to the females; otherwise the close inbreeding may result in a higher than normal incidence of deformities.

Selling fertile eggs and day-old chicks

Many breeders of pure breeds sell fertile eggs and day-old chicks. Delivery of these is more straightforward than for older birds, which are covered by the stringent transport and delivery of livestock regulations. Day-old chicks have the remnants of the yolk in their abdomens and so do not require food for 48 hours.

A frequent complaint from purchasers of

day-olds is that there are too many males amongst them, even though they may have paid extra to have sexed chicks. It is cheaper to buy them 'as hatched' (AH). Sexing day-old chicks is a difficult thing to do, requiring a great deal of experience. It is not a good idea to attempt it without practical training because it is easy to damage the chicks. It is better not to sex them and accept a lower rate of payment, rather than risk hurting them and also having complaints from customers.

Another source of contention is for eggs or day-olds to turn out to be different from the breeds they are sold as. It is essential to keep different breeds separate if their progeny are to be sold. Customers have a right to expect what they have paid for, as well as breed details of the parents. If they are sold cross-strains as pure-bred stock, they will not be happy with the vendor, and who can blame them?

If older birds are being sold, it may be better to arrange for them to be collected, or to meet the buyer at a halfway point. There are specialist delivery services available, but they are expensive for small numbers of birds.

Cockerels

It is a fact of life that most cockerels are unwanted, although some can be raised for the table. The point is obvious in relation to the raising of laying stock, and many of the larger enterprises either dispose of the day-old males or sell them to zoos which use the dead chicks as food for inmates such as large birds of prey. Where they are disposed of on site, there are humane considerations to be borne in mind, and the Ministry of Agriculture provides guidelines in this respect. A leaflet is available from them, on request. The Humane Slaughter Association's guide *Practical Slaughter of Poultry*, referred to earlier in the book, also contains helpful information.

CHAPTER 12 **Showing**

Many who visit poultry shows and see the fowl in a spick and span condition, little know of the training, washing and conditioning that these specimens receive before they are shipped to the show room.

Michael K. Boyer, 1908

All poultry keepers have, at some time, had a particularly impressive looking bird that they felt was worthy of a prize, even if they never did anything about it. Most people do not show their birds, but what if they decide that they would like to do so? Where do they begin?

It is often difficult to choose which breed to keep for show purposes, but whichever is chosen, it has to be a pure breed that is recognised by the Poultry Club of Great Britain. No matter how beautiful a cross-bred or hybrid bird may be, it will not be accepted

Viewing the breeds at the National Poultry Championship Show.

in any of the classes run under the auspices of the PCGB.

There are several centres where pure breeds are kept, and which are open to the public. The most well known in Britain are Wernlas, the Domestic Fowl Trust and the Woodchurch Rare Breeds Centre. Poultry shows and agricultural shows with poultry tents are also worth visiting in order to see a range of breeds. The premier poultry shows are the National Poultry Championship Show and the Poultry Federation Championship Show, both of which are held annually, in December. In addition to these, there are many regional and local shows, all held under the auspices of the PCGB. A good cross-section of breeds can be seen at venues such as these, and it is an excellent opportunity to meet breeders and talk to them about their particular breeds.

Before buying any birds of the chosen breed, it is a good idea to establish what is the ideal for that breed. The best way of doing this is to join the appropriate breed society and obtain a copy of the relevant standards. These list the characteristics of the male and female, describing the ideal attributes of stance, feathering, colouring and so on. There is a system of points allocated for different characteristics, with the total points adding up to 100. A bird that is exhibited at a show will be given a score based on its ability to measure up to the standard. The higher the score, the better. Standards for all the breeds are in the book *British Poultry Standards* published by the Poultry Club of Great Britain and Butterworths. An example of a standard, that of the Maran, is given in Figure 12.1.

After this, it should be possible to purchase the best possible stock. Someone entering the show world for the first time, however, would be lucky to do this straight away. Most breeders, understandably, keep their best stock for breeding and sell the progeny. It may be necessary to improve the birds over a period of time, breeding and selecting the best ones. It is also a good idea to buy stock

for breeding from two different sources so that they are not related. Some breeds that are few in number are probably related already.

It is a fair question, perhaps, to ask why showing is necessary. Apart from the obvious enjoyment that many people receive from it, it does have other aspects. The competitions provide a focus for the breeds, with the competitive element undoubtedly ensuring that the show standards are maintained. It is also fair to say that, although utility characteristics have declined, some breeds would have become extinct had it not been for the continuing interest of the poultry fanciers. When a breed does become extinct, it means that its genetic potential has gone for ever.

Housing

Any of the various houses available for small poultry keepers will be suitable for show breeds. It is a matter of individual sites and preferences. Some prefer to use a house with run for a male and several females of the same breed. Each breed kept should have its own house and run. Some people prefer an aviary system, in which an outbuilding is adapted to take several individual pens. Some fanciers keep their birds inside all the time, fearful that outdoor conditions will have a detrimental effect on their birds, such as the yellowing of white feathered birds by the sun. While this is a problem, it is obviously more humane to compromise and let birds have access to a shaded run. If a bird has to spend a day in a show cage at a poultry show, the least it should expect when it arrives home is a house with an adequate exercise run.

Whatever the housing facilities, they should be equipped with a perch, feeder, drinker and, where females are housed, a nest box. Once breeding starts, other housing will be required. An incubation room or indoor area is a good idea, and chicks will

Origin: France
Classification: Heavy
Egg Colour: Dark Brown
General characteristics: Male
Carriage: Active, compact and graceful.
Type: Body of medium length with good width & depth throughout; front broad, full & deep. Breast long, well fleshed, of good width & without keel lines. Tail well carried, high.
Head: Refined. Beak deep & of medium size. Eyes large & prominent; pupil large & defined. Comb single, medium size, straight, erect, with five to seven serrations and of fine texture. Face smooth. Wattles of medium size & fine texture. Neck: medium length, not too profusely feathered.
Legs and Feet: Legs of medium length, wide apart & good quality bone. Thighs well fleshed, but not heavy in bone. Shanks clean & unfeathered. Toes, four, well spread & straight.
Plumage: Fairly tight & of silky texture generally.
Handling: Firm, as befits a table breed. Flesh white, and skin of fine texture.
General characteristics: Female
Similar to those of the Male, allowing for the natural sexual differences. Table & laying qualities to be taken carefully into account jointly.
Most popular colour: Dark Cuckoo.
Male and Female plumage: Cuckoo throughout, each feather barred across with bands of blue-black. A lighter shaded neck in both Male and Female, and also back in Male, is permissible if definitely barred. Cuckoo throughout is ideal, as even as possible.

There are also Golden Cuckoo, Silver Cuckoo, and Black Marans.
In both sexes and all colours: Beak, white or horn. Eyes red or bright orange preferred. Comb, face, wattles & ear-lobes red. Legs and feet white.
Weights: Cock 3.6 kg (8 lb); Cockerel 3.2 kg (7 lb) Hen 3.2 kg (7 lb); Pullet 2.7 kg (6 lb)
Scale of points:

Type, carriage and table merits (to include type of breast and fleshing, also quality of flesh)	40
Size and quality	20
Colour and markings	15
Head	10
Condition	10
Legs and Feet	5
	100

Serious defects:
Feathered shanks. General coarseness. Lack of activity. Superfine bone. Any points against utility or reproductive values.
Defects (for which a bird may be passed)
Deformities, crooked breast bone, other than four toes, etc.
Maran Bantams should be true miniatures of their large fowl counterparts.
No black Bantam is standardized.
Weights:
Cock 910 g (32 oz) Cockerel 790 g (28 oz) Hen 790 g (28 oz) Pullet 680 g (24 oz)

Source: The Maran Club. Also published in the article 'The Maran', Fred Hams, Country Garden & Smallholding, July 1996.

Figure 12.1   The Maran British Poultry Standard.

need protected brooding conditions. Growing pullets and cockerels will also require separate accommodation from the older birds. The trouble with breeding, as they say, is that it goes on and on.

Breeding

The elements of breeding were covered in Chapter 11. Few people within the pure breed sector breed from their own birds in order to carry out selection and improvement. Yet it can be done. Ideally there should be trap-nests, such as the one illustrated on page 121, so that it is possible to determine which hens lay which eggs. If every breeder of pure fowl made an effort to select and breed for vigour and productive capabilities, as well as standards considerations, it would be an important stepping stone towards ensuring the future of traditional breeds. Too many are in-bred at present, with an increasing incidence of fertility loss and birth defects. It is becoming increasingly necessary to introduce new

Most country shows have a poultry exhibit.

blood, and it has to be faced that this can only come from the hybrid sector. Introducing hybrid vigour is a positive step in the production and standardisation of new strains and breeds. It is quite a challenge.

The principle of using a trap-nest in order to establish who are the best layers was covered in Chapter 11. Other factors that are important are freedom from defects, robust constitution and the appropriate standards. Poor birds should not be selected for breeding. Those that are suitable should have leg rings for identification. Rings are a useful means of identifying birds at shows, as it is not unknown for a judge to put a bird back in the wrong cage. Birds can also be stolen. A numbered leg ring that cannot be removed from the bird is a positive identification. The Poultry Club of Great Britain has a Ringing Scheme by which rings are supplied to members and the details are registered with them. In the event of selling

a ringed bird, notification can then be made to the Club.

There is a procedure within the PCGB for the inclusion of new breeds or varieties. If you are successful in your selective breeding and produce a new breed, you can apply for registration of the proposed name. Its genetic characteristics must be fixed, in the sense that it will produce offspring like itself, and must be sufficiently distinctive from existing breeds. Evidence of breeding lines for at least two generations must be supplied, as well as details of the proposed standard with a signed declaration. The proposed breed should also have been exhibited in non-standardised classes at a show approved by the PCGB. This application will be considered, with submissions also made to the breed club on which the new strain was based, or to the Rare Breeds Society, if applicable. Eventually the proposal is voted on by the Council.[1]

Feeding

Breeders' compound rations are available for breeding birds, but often in large quantities only. As feeds have a limited shelf-life, a large batch may go off before it can be used. If it is available in smaller quantities, it is worth using it for any birds that are to provide fertile eggs for incubation. In most cases, it is appropriate to feed show birds a similar ration to that given to utility fowl. A high protein layers' mash or pellets together with grain is suitable. Grit and crushed oystershell should be available, as well as fresh water at all times. In Chapter 11 the importance of balanced feeding for breeding birds was stressed. Unless they have adequate minerals and vitamins, the chicks may have problems, either in the shell or at hatching. A mineral and vitamin supplement would ensure that these do not occur.

All the general advice given on the feeding of utility stock applies to show birds, and the need for grit for the efficient working of the gizzard should not be overlooked if grain is given.

Training

An indoor area or penning room is useful for a range of general show preparations. It is important that the birds become accustomed to being handled; it would not do for them to be so flighty that they cannot stand still or that they panic when the judge comes to look at them.

Talk to them while handling them, as this has a calming effect, and they will very quickly become used to their handler. It is also important that they become accustomed to being put in and taken out of show cages. This should be practised particularly before a show, remembering to talk to the bird all the time.

When taking a bird out of its cage, open the door slowly and slowly insert one hand underneath the bird, while confining the wings with the other hand. Avoid any sudden jerking movements, and there should be no problems. A good way to teach standing still is to place the chicken on a level surface and stroke it along the back with a short piece of bamboo. This will accustom it to the judge's examination.

The bamboo stick can also be used to train the chicken to stand in the way appropriate to the particular breed. For example, the Indian Game is required to adopt a stance with the legs wide apart. A stick is not essential, of course, and the bird can be stroked with the hand, but it is important to teach the bird to stand in the right way for its type. The standards for the appropriate breed will indicate how this should be.

Show preparations

The emphasis should be on helping the bird to make the best of itself. White and other light-coloured chickens may need a wash in warm water and baby shampoo, which is gentle but effective and will not cause distress if it gets in their eyes. Detergent is not advisable. Three plastic basins are best, two with soapy water and one with clear rinsing water.

If the bird's feet are particularly dirty as a result of its scratching activities, wash them first, using an old toothbrush and soapy water to remove caked dirt. Clean the scales of the legs in this way as well. Next keeping a firm hold on the bird so that it cannot flap its wings, lower it into the second basin of soapy water and begin to work the lather into the plumage. Use a small piece of sponge to clean the face, and take care to avoid getting water in the eyes. Gently squeeze out the surplus lather from the feathers and then transfer the bird to the bowl of rinsing water.

Towel the bird dry, making sure that the feathers are not dragged roughly. Smooth the feathers into place and put the bird in a warm place to dry, or use a hair dryer on the

Dark Brahma male taking part in a poultry show.

cool setting to dry quickly. Use the dryer in such a way that it blows the feathers into position, rather than against the natural line. Whichever method is used, it is important not to subject the bird to sudden fluctuations of temperature or cold draughts at this stage.

Dark birds do not usually require washing (unless they suddenly become caked with mud), although the legs and feet should be washed like those of white birds. The birds are normally brushed to remove any extraneous matter. A piece of silk is useful for bringing up a sheen to the feathers. This is rubbed along the plumage, following the natural line of the feathers.

It occasionally happens that a small feather sticks out in a way that detracts from the bird's appearance. As long as it is not one of the prime or secondary feathers and its absence will not be obvious, there is no reason why it should not be gently removed. Although nothing should be removed from or added to show birds before judging, this does not apply to the occasional loose down feather.

Travelling and exhibiting

The Welfare of Animals During Transport Order, 1994, must be complied with when travelling with animals. As far as the show person is concerned, it means having proper travelling containers that are adequately constructed and ventilated so that they are suitable for providing comfortable and stress-free conditions. There is also a requirement that no bird should travel more than 12 hours without feed and water. There are specialist suppliers of show cages for housing chickens while they are on view, as well as purpose-made travelling boxes. Show organisers will be governed by The Diseases of Poultry Order, 1995, in as far as it is necessary to clean and disinfect the floor, tables, pens and fittings, as well as transporting vehicles, before and after use.

It is a good idea to go and see how poultry shows operate before starting to exhibit yourself. It is interesting, not only to see the birds, but also how the judges examine them and what comments are made. Taking a look at the winners in their cages, complete with rosettes, gives an indication of what is approved of. If you are a member of a breed club, helping at a show, possibly as a steward, is also a good way of gaining valuable experience and possible insights. A steward will need to wear a white coat.

The first step to taking part in a show is to obtain the show schedule from the show organiser. This will have details of the classes available and the conditions of entry. There are often juvenile classes for the under-16s, so children in the family may also like to take part. There are also egg classes, so if you have eggs to be proud of, they can make an appearance. (Britain is one of the few countries in the world to have egg classes in poultry shows.)

Decide on the relevant classes, fill in the form and post it with the relevant fees. There is a cut-off date by which time entries must be received.

On the day of the show, ensure that the potential champion is comfortably in the travelling box and that feed, a bottle of water and a clip-on drinker are packed. Take some wood shavings in a box in case there should be a shortage at the show. Any last minute grooming aids can be taken along, such as a cloth or piece of silk. It goes without saying that a sick bird should not be taken.

In the vehicle, ensure that the travelling box is held securely in place so that it cannot move about. Set off to arrive reasonably early, following the show organiser's instructions about which entrance to go to, where to park and where to present the entry for inspection.

Ensure that the bird is happily ensconced in its pen, then stand back. When the judging takes place, owners of bird entrants should be well clear and not attempt to speak to the judge.

When the verdict is made, all is revealed. Your champion was not even *commended*? Never mind, there is always the next time!

References

1. Registration of New Breeds, *Poultry Club Year Book*, 1996.

The winner!

CHAPTER 13 Marketing

If you can't identify your customers, forget it!

Production is one thing, marketing is quite another. Marketing is the development of a product to fill a demand. If a distributor is used, the onus for marketing will be on him and the producer can concentrate on production. This is the simplest course of action for anyone producing free-range eggs because the distributor will collect, grade, pack and deliver the eggs to the shops.

A good starting point is to visit several large supermarkets and investigate their free-range eggs. How are they packaged? Are they attractively presented? What price are they? Who distributes them? Supermarkets will not take eggs direct from a producer, but their head offices will tell you who their distributors are, and how to get in touch with them.

Distributors will normally provide information about the type of eggs required and will frequently sell speciality eggs in attractive packaging. It must be emphasised, however, that there must be sufficient numbers of eggs produced on a regular basis before a packer will agree to a contract. This means having more than one flock of several thousand birds to ensure that there is no dip in production.

The point was made earlier in the book that as far as table birds are concerned, it is best to concentrate on organic production because this is where a shortage lies. Contact UKROFS or one of the organic organisations such as the Soil Association who are registered with it. Not only will they provide details of organic production and how to become a registered producer, but they will also offer advice on distribution.

For smaller producers, it is more appropriate to do their own grading, packing and marketing, particularly if a farm shop is available on-site. Selling eggs is easier than table birds in a farm shop situation. The point was made in Chapter 5 that planning permission is more difficult to obtain where there is an element of food processing, as in the preparation of birds for the table. The Egg Marketing Inspector at the nearest DEFRA office will supply a copy of the European Union rules which govern egg sales, while DEFRA and the local Environmental Health department will have details of legislation which applies to food processing. Further details are given in the chapter on eggs.

Where the small producer is doing his own distribution, it is crucial to bear in mind the basic principles of marketing.

The marketing plan

The strategy is outlined in Figure 13.1 and involves identifying the customer, establishing the product, getting the price and the site right, and then promoting the product.

Identifying the customer

I made the point earlier in the book that there

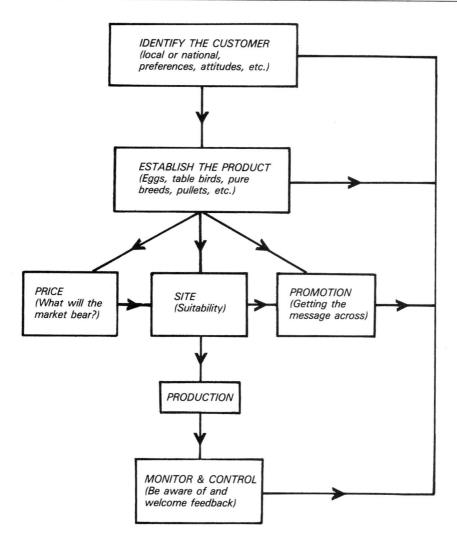

Figure 13.1   The marketing plan.

are organisations that offer a consultancy service which includes the necessary market research for a particular area. Attending a course on free-range production of the type run by an agricultural college is highly recommended too. There is also a great deal that the individual can do to identify the potential customer and his attitudes. On a local basis, find out from delicatessens, restaurants, butchers, hotels and guest houses whether they are interested in buying. Speak to local consumer groups, Women's Institutes and environmental groups, and listen to what they have to say. If you are already in production in a small way, try taking a market stall, not only to establish buying patterns, but also to have the opportunity of talking to people. A great deal can be learnt by good listening.

Once you are satisfied that there is a demand, the decision may be whether to concentrate on local, farm-shop sales, or to cater for national demand via a distributor. The two activities are not mutually exclusive, but the former will obviously require far less capital expenditure than the latter. Most free-range producers are small, catering for local sales, but most free-range eggs come from large enterprises selling to chain retail outlets.

Establishing the product

Establishing the specific product will be largely determined by the initial identification of the customer. There is not much point in launching into egg production if all the market research has indicated that there is a local glut of free-range eggs in the supermarkets, and what customers are looking for is a source of free-range table poultry. What is needed now is a more detailed examination of the specific product. For example, if there is a demand for free-range eggs, is it better to concentrate on the production of speciality ones such as those that are very deep brown? Again, some test marketing in the form of initial small-scale production and local sales may indicate the path to take.

The price

The price of the product is crucial. Mr Micawber, in *David Copperfield*, sums up the recipe for economic happiness as 'Income £1, expenses 19s 6d' - but the aim should not be just to cover costs and have a bit of profit at the end. The positive approach is to ask the question, 'What price will the market bear?' This is the starting point with any quality product, especially where production costs are higher anyway. Check the prices that are being asked in supermarkets for free-range and organic eggs, as well as free-range and organic table birds. Visit other farm shops to see what they are charging.

Keeping in touch with market information is vital and there are several organisations which issue regular details of price changes. They include United Kingdom Egg Producers' Association (UKEPA) and British Free Range Egg Producers' Association (BFREPA). DEFRA also has a weekly statistical bulletin on the poultry industry. Local National Farmers' Union (NFU) offices produce a weekly update of pricing information available to their members, and the poultry press also publishes regular updates.

The site

The site where production is taking place is crucial if this is also to be the area of sales distribution, such as a farm shop. If an expensive, quality product is being produced, it is not likely to achieve its maximum sales potential if it is in an isolated area. A successful farm shop needs to be close to a fairly highly populated, reasonably affluent and mobile centre of population. Proximity to an urban area may not necessarily lead to successful sales if it is also a region of high unemployment. In such areas, the cost factor is likely to take precedence over considerations of quality.

The ideal road for a sufficiently high volume of calling customers at a farm shop is a two-way 'A' road which carries a substantial proportion of commuter traffic to a nearby city. A straight road, offering plenty of areas for warning notices of the shop ahead, enables drivers to see clearly, and to have sufficient time to respond. There should be plenty of turning and parking space, without the danger of causing a traffic hazard. Any changes, such as the diversion of traffic, could be disastrous to an enterprise. Direct gate sales are the 'bread and butter' of many producers, providing an effective security against the fluctuations of packers' prices, so it is essential to check on future transport plans in the area before going into business.

Any site where the public calls should, of course, have comprehensive insurance cover.

Promotion

Producing a marketable product is one thing, but promoting it is another. The message has to be got across to customers that here is something which is clean, wholesome, environmentally and ethically sound, and definitely good for you. Keep the marketing advantages of free-range eggs and table birds clear in your mind. The point has already been made that most people who buy eggs, for example, buy battery ones because they are cheaper. The marketing of free-range eggs should be firmly targeted at those consumers who demand a 'quality' product. The perception of quality in this case is of a product that is natural and healthy and is produced with a concern for humanitarian and welfare issues. This distinction would be even more apparent if intensive producers were required by law to label their eggs as 'battery cage-produced' . Unfortunately, they are not required to do so, nor are they prohibited from describing their eggs in such terms as 'country produce fresh from the farm', when the reality is that they are 'fresh from the industrial unit of the battery'.

The appeal of cheerful posters, leaflets and signs is not to be underestimated. Even a basic desk-top publishing programme in a home computer can produce eye-catching visual material. There is usually a local print shop in most towns where posters can be enlarged or printed with spot colour at a reasonable price. Many feed companies also provide material that can be used. Good packaging and labelling are also essential.

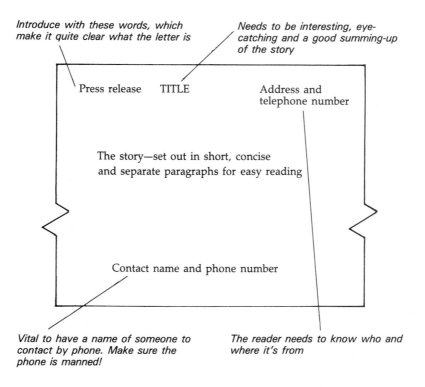

Figure 13.2   How to produce a press release.

(See the eggs chapter for further details.)

If sales are mainly local, then local publicity is highly desirable. Sending a press release to the local newspapers and radio and television stations about your activities is one way of achieving this. If the enterprise includes a farm shop, some kind of attraction could be added. If poultry and pet foods are being sold, as referred to below, a small children's pets display might be appropriate. An attractive, welcoming place is far more likely to attract customers again if they are made to feel that their visit has been worthwhile. One free-range producer who ran a small survey among his customers to discover what they liked best was surprised to find that a frequent comment was 'the nice, clean toilets'!

The free-range smallholding

Figure 13.3 shows a smallholding of 3 hectares, specialising in free-range eggs. There is a farm shop and every effort is made to present the eggs attractively. This entails having a range of different coloured eggs whose novelty value appeals to customers. Speckledy, Maran and Welsummers produce the dark brown eggs, while a few Araucanas are kept for their greenish ones. The latter are particularly popular at Easter and Halloween. A few Leghorns are kept for their pure white eggs, just to provide contrast, while the bulk of the pinkish brown eggs are produced by a brown egg hybrid such as Lohmann Brown, Hisex Ranger or ISA Brown.

As there is a large kitchen garden at the smallholding, some produce is offered for sale in the shop, with the emphasis on seasonal availability. One of the recent bestsellers has been pumpkins for Halloween! Home-made pickles, chutneys and jams are also on sale.

The small flock of sheep is a traditional breed with coloured wool, such as Shetlands. They graze with the hens, keeping the grass down to an acceptable length. Their fleeces are processed to produce wool for spinning, as well as the end-products of various wool crafts. This aspect has been instrumental in several school parties coming to visit the site, so a range of postcards has been introduced in the shop. It is obviously a family-run enterprise; otherwise the range of activities would not be possible.

This particular smallholding is on a main road, with easy access and plenty of parking and turning space. There is a large barn for feed storage, so a range of poultry and pet feeds and supplies are also stocked for sale. This provides a regular income which is a useful supplement to egg and other sales.

There are many such enterprises, but the successful ones are those which have done their homework. It is a truism to say that the market comes before the chicken or the egg but, like all truisms, it contains a fundamental truth. Distribution should be researched and established before production begins.

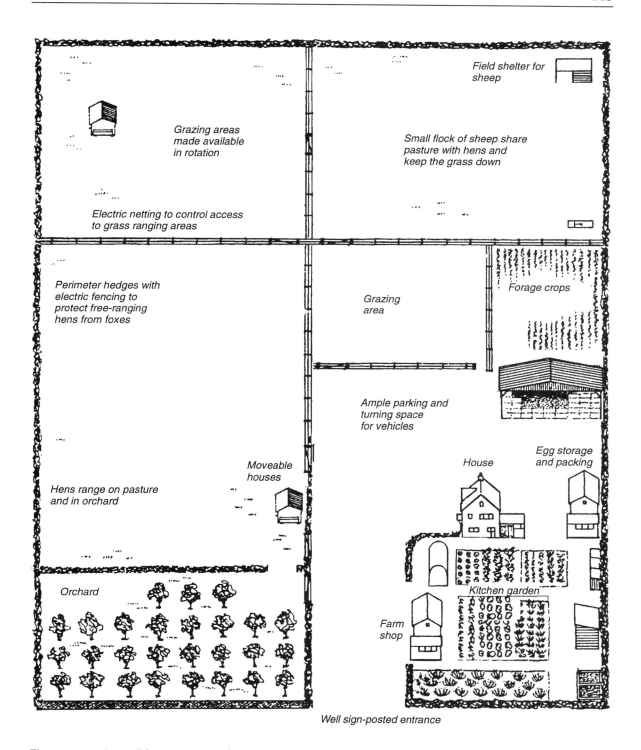

Field shelter for sheep

Grazing areas made available in rotation

Small flock of sheep share pasture with hens and keep the grass down

Electric netting to control access to grass ranging areas

Perimeter hedges with electric fencing to protect free-ranging hens from foxes

Forage crops

Grazing area

Ample parking and turning space for vehicles

Egg storage and packing

House

Moveable houses

Hens range on pasture and in orchard

Orchard

Kitchen garden

Farm shop

Well sign-posted entrance

Figure 13.3   A small free-range egg farm with a farm shop.

CHAPTER 14

Common Problems and Diseases

Prevention is better than cure.

The aim with all poultry is to avoid problems before they happen. Most hybrid pullets destined for the free-range sector will have been vaccinated against Marek's disease, Newcastle disease, infectious bronchitis, gumboro disease, endemic tremors and egg drop syndrome. If birds are raised organically, as egg producers or for the table, the Soil Association's Symbol Standards do not allow the vaccination of poultry in this way.

Hybrid pullets will also come from a breeding flock that is regularly tested for the presence of salmonella, and where the parent birds are free of inheritable diseases.

Chickens need to be provided with clean, well-managed housing and grass, good food and clean water. With these, and regular action taken to deter parasites, vermin and predators, they are unlikely to develop anything serious.

The three 'Ds' to beware of are dirt, dust and droppings. All three harbour disease organisms and need to be dealt with on a regular basis, in the house and on the grass. Frequent change of grazing is necessary in order to avoid a possible build-up of infection, unless the flock density in relation to the amount of land is low. Long grasses are hiding places for a range of infective organisms and keeping grass cut down is a positive way of minimising the risks of infection. Cutting also allows sunlight, which is a great cleanser, to have access to shorter grasses. Pasture which is being rested will benefit from an application of lime which not only helps to 'sweeten' the soil, but contributes to breaking the life cycle of parasitic worms which may have infected the site.

Freedom Food producers are required to have a veterinary health plan drawn up and regularly updated by a visiting veterinary surgeon. Flock performance needs to be continually monitored for evidence of production disorders or signs of disease. If performance falls below the tolerance limits specified in the veterinary plan, the vet must be informed and the plan revised to include a remedy programme. Particular conditions to look out for are egg peritonitis, cannibalism, significant feather loss and red mite infestation.

Commercial producers are required to keep a record of all medicines administered to their poultry, as well as to observe any withdrawal periods. Some products are available from licensed stockists, while others are available only from vets.

Many problems are not necessarily related to disease, but can still be a nuisance. It is worth looking at the most common ones.

Common problems

Floor-laid eggs

This could be for a number of reasons. Too

many birds competing for too few nests is one. The optimum ratios are:

Small house (6 to 12 birds)   1 nest box per 3 birds
Large house   1 nest box per 5 birds

Floor-laid eggs are much more common in large houses than in small ones. Lighting that creates harsh shadows can confuse the birds. Some large houses are very gloomy because the light is kept dim to prevent aggression. Houses with natural daylight are preferable. Nest boxes need to be in a slightly darker area to make them attractive as nesting places, but they should be visible.

The most common reason for floor-laid eggs is that the chicken is confused as to where the nest box is, or that there are too many obstacles in the way. Obstacles are not necessarily permanent fixtures, as they can be other birds. It can simply be too much trouble for the chicken to go a long distance through a crowd so she lays on the floor instead.

Reference was made on page 63 to the producer who was advised to raise the nest boxes off the floor so that the hens could see them more easily. Some producers use an electrified wire to prevent floor-laying. This cannot be justified on welfare grounds and hopefully the practice will be discontinued.

Litter can sometimes be attractive as a nesting area. Do not let wood shavings accumulate on the floor. Where there is a scratching area, use sand and ensure that the area is well lit. If necessary, pot eggs can be used to train new pullets to lay in the nest boxes.

An interesting study some years ago indicated that painting the nest boxes grey reduced the incidence of floor-laid eggs. The trial was conducted in a large deep litter house where 417,168 eggs were laid. Most of them were in the nest boxes, but 7.2 per cent were floor laid. When the nest boxes were painted, the following results were obtained:

Percentage of floor-laid eggs with different coloured nest boxes

Red	3.5
Green	2.5
Grey	0.8
Unpainted	3.0

Source: Ptitsevodstvo study, Russia, 1992.

Dirty eggs

This is fairly easily rectified. Make sure that the nest box liner is clean and that there is an area outside the house where the chickens can wipe their feet before entering. A covered area which is dry helps a great deal.

Egg eating

Every poultry keeper has come across the bird that eats eggs. There are several causes. Boredom in a small run is one, and the provision of suspended kitchen garden greens is important in deterring it. A period of hot weather, lack of water or calcium, or insufficient food may also cause it. Collection of the eggs at infrequent intervals may provide an accidentally cracked egg. This is then eaten, leaving a taste for others in the future. A habit can develop very quickly.

Make sure that the birds have enough crushed oystershell. The lights in a large house can be dimmed slightly while, in a small house, a vertically slit plastic curtain can be fitted to the nest box entrance to provide darker conditions.

If it is possible to identify the culprit, she can be put in another run for a couple of days. There is nothing like a change of impressions to break habits. She should still be in sight of the other birds; otherwise they may attack her as a stranger when she is returned to the flock.

If this does not work, try putting a mustard egg in the nest. Take an egg and crack it gently so that the contents can be removed. Enlarge the hole slightly and fill it with mustard. Seal it with sellotape and put in the nestbox, tape side down. If that does not work, nothing will.

Where rollaway nest boxes are used, this problem is avoided, as the eggs are generally out of reach of the hens.

Feather pecking

One cause of this is over-crowding, which is why most hybrids are routinely beak-trimmed. In large houses the problem can still occur and reducing the light intensity is normally effective. This must be done gradually so as not to disrupt egg laying.

In low density flocks feather pecking is often connected with lice and mites. If a bird is pecking at her own feathers she may inadvertently draw blood, thus inviting pecks from others. Ensure that the feeding is adequate. If there is insufficient protein, this could be a contributory factor. Give an ordinary free-range layers' ration, avoiding the type that is geared to maximum production as part of phased feeding.

Some batches of birds are undoubtedly worse than others, reinforcing my belief that there may also be a genetic predisposition towards aggression.

Vent pecking

This is a variation on the previous problem, but is more serious because it can lead to the death of a badly pecked bird. Again, it is more common in large flocks, particularly where rich feeding leads to the production of over-sized eggs. (See page 84 for further information.) This can lead to tearing of the vent and prolapse, in which the lower part of the oviduct turns inside out, emerging from the vent. Vent pecking can also happen when pullets come into lay too early.

Reducing stress conditions by introducing a more 'gentle' feeding regime and by reducing flock density helps to prevent vent pecking. Large houses rely on reducing the light intensity.

Cannibalism

This is feather and vent pecking carried to extremes. To avoid it, follow the advice given earlier.

Unseasonal moulting

Moulting in the summer is normal. Prolonged or unseasonal moulting is abnormal and should be investigated. First check that there is not an attack of depluming mites. If so, treat with a proprietary dusting powder. It is most likely that moulting has been brought on by stress. In large flocks, stress is often a combination of high-energy feeds and lack of exercise. Heat stress in a poorly ventilated house can also be a contributory factor. Again, using an ordinary layers' ration and grain and encouraging wider ranging are less stressful to the system. This question was examined in the chapter on feeding.

Stress leading to abnormal feather loss in a domestic flock can sometimes be traced to factors as varied as the birds being fed on nothing but kitchen scraps, to being worried by a vicious family dog. I have also heard of one small poultry keeper who fed the hens nothing but oats, believing that this was what they required. Oats are extremely 'heating' and a surplus will almost certainly cause moulting.

Convict's foot

This is not a name that you are likely to come across in a poultry book because it is one that I have coined. The condition happens when a hen scratches about in an area where there is straw. An accumulation of litter gradually forms a hard, solid ball around one of the claws until the hen looks like a convict with a wattle and daub ball and chain.

It is not possible to remove the ball without first soaking it well; otherwise the claw may be damaged. It may require several soakings to get it all off because it is so hard and strong. The use of wattle and daub as a traditional building material is understandable.

Egg with blood on the shell

Young pullets coming into lay early sometimes strain to lay and tear the vent. Blood is also caused by over-sized eggs as a result of feeding 'over-rich' feeds, as referred to above. Also avoid giving light too early.

Egg with blood spots

This is usually the result of blood escaping from the ovarian follicle and becoming embedded in the albumen. It is sometimes caused by stress or a sudden shock.

Soft-shelled egg

The first egg a pullet lays before the egg laying system gets into its stride may be soft-shelled. The problem is usually temporary. If not, make sure that crushed oystershell is made available. Occasionally an egg may be completely devoid of a shell.

The condition can also occur in hens well into lay. My own observations are that sudden changes in the weather can be a contributory factor. I have known hens that have been soaked in the rain to produce such eggs, and then the next day be back to normal.

Wind egg

This is the name given to an egg without a yolk. Again it is often a pullet's first egg and the problem is not likely to recur. In an older bird it is likely to be the result of a sudden shock.

Egg with green yolk

This is what happens when a free-ranging hen finds shepherd's purse plants or some acorns, both of which turn the yolk green.

Egg with pale yolk

In winter, when the grass is not growing, the yolks are pale. Feeds with grass meal and maize improve the colour, but there is no need to resort to artificial yolk pigments. Nothing is wrong with pale yolks except perhaps customer preference.

Fertile egg

The cock should not be running with the hens if eggs are for sale.

Double-yolked egg

Two yolks have been released into the oviduct at the same time and have subsequently been enclosed by a shell. They are fairly common in large eggs.

Egg with watery albumen

Ammonia from badly managed litter can cause this, or an illness such as infectious bronchitis. Sometimes it may be a reaction to vaccination. A multi-vitamin supplement in the drinking water may improve matters because it is a good way of dealing with the stress that may have caused the problem. The condition is more common in hot weather, and tends to be found in older hens.

Middle-banded egg

Sometimes an egg may be temporarily halted within the system, usually as a result of shock or sudden disruption. The result is a band or ridge around the centre of the egg. It is nothing to worry about, but ensure that the birds are not being disturbed or frightened in some way.

Misshapen egg

These differ from the middle-banded egg in having a range of distortions, including soft ends and uneven or ribbed surfaces. If the eggs appear regularly, they may be from an old hen. If not, diseases such as infectious bronchitis or egg drop syndrome should be suspected. See the A-Z guide later in this chapter.

Pale eggs
Loss of shell colour is frequently associated with a period of bright sunshine when hens may also have lost feathers from their backs. (See page 148: unseasonal moulting.)

Diagnosing a problem

Commercial producers registered with Freedom Food will be working in conjunction with a vet. The following is therefore for the small or domestic flock owner. It is not intended to replace veterinary diagnosis, but to provide some guidelines as to when veterinary advice should be sought.

A hunched-up stance is typical of an ill bird. The first thing to do is to separate it from the flock and put it in a small 'hospital' shelter and run, separated from, but in view of, the others, so that if it goes back, it will not be an outsider.

Examine the chicken and note its appearance. Try to establish whether the problem is an external or an internal one. External problems include wounds, cuts, bites and stings, and are generally easy to detect. Check the head, body, wings and legs for signs of damage, cuts, bruises, swellings and, of course, little grey parasites scuttling along the skin. If lice are visible, treat the hen with a dusting of proprietary powder or spray. It is essential to keep chickens free of external parasites. They can be extremely debilitating, causing anaemia and even death if the infestation is a chronic one.

Wounds and cuts can be cleansed with cotton wool, warm water and household disinfectant or hydrogen peroxide. If there are signs of lameness, it may be a previously undetected cut which has healed over a wound containing pus. This is a condition called bumblefoot (information is given in the A–Z listing further on). If there are white encrustations on the legs, it is scaly leg caused by a mite. Reference was made to this

in the summer management section on page 99. Again, this is a condition that should be cleared up without delay.

If the problem is not an external one, then something is going on inside. Feel the crop area to detect if there is any compaction. If the hen has had access to long pieces of loose grass there may be a blockage. Sometimes gentle kneading can loosen it, but a few drops of cooking oil trickled into the beak is more effective. It may not do any good, but it will not do any harm either, and I have in the past saved birds with impacted crops in this way.

If the crop is not blocked, does it feel like a bag of water? A dose of Epsom salts may sort things out if she has sour crop. Again it will do no harm and may do some good.

If the crop feels normal but the hen is obviously off-colour, the third remedy in the cupboard is cider vinegar. Add a little to warm water so that it is not too acidic to the taste, and trickle it into her beak.

If the problem is none of these, is it possible that she is egg-bound? This can happen. If a chicken is seen going into the nestbox and then emerging, only to go back a short time later, there may be a problem. It is best to leave her for a while to see if the egg does eventually come. If not, she needs help.

If the vent is bleeding slightly and she is straining, the vent will pulsate to the extent that it might even be possible to see the egg inside. Great care is needed, because if the egg breaks inside her she will probably die of shock and infection. The only thing to do is to gently apply some Vaseline to the vent. Get a bowl of boiling water and hold her vent just above it so that the rising steam bathes the vent. This can have the effect of relaxing the muscles to the extent that the egg pops out.

If the hen is not egg-bound, check her for the possibility of internal parasitic worms. Is her comb anaemic? Is she excessively thin? Does the breast bone feel sharp, with little flesh on either side of it? If she needs worming, Flubenvet can be administered in the food, as detailed on page 99. Flubenvet can be acquired only from a licensed stockist.

If it is not clear what is wrong with her, the best thing to do is just to give a dose of cider vinegar in water. Leave her in the protected pen with an ordinary drinker and a little layers' mash overnight. You have done all you can. In the morning, she may surprise you by having made a complete recovery.

If she does not recover, keep a careful watch on the rest of the birds. If others begin to sicken, call the vet!

Salmonella

There is a requirement for those with a breeding flock of 250 birds or more to test their breeders for salmonella.

One of the problems with some salmonella organisms is that they can be carried by apparently healthy birds and passed directly into the egg, no matter how clean the environment may be.

Salmonella enteritidis and *S. typhimurium* are examples of this. There is some evidence that increasing the acidic level of the feed inhibits salmonella in chickens. These are feed additives available which are made up of organic acids. They are formulated into a granular material which is mixed with the normal feed, and in this form they inhibit salmonella and other pathogenic organisms. The traditional practice of adding a little cider vinegar to the feed or water, every so often, would appear to have a sound scientific basis.

It is now possible to vaccinate against both forms of salmonella mentioned above. Eggs that are marketed under the Lion brand are from hens vaccinated in this way so that the chances of consumers becoming ill are minimised. Vaccination does not eliminate the risk entirely, hence the continuing advice that consumers should store their eggs in a refrigerator and not give soft-boiled eggs to those who are young, old, ill or pregnant. (Hard-boiling eliminates the risk.) However, it has been estimated that vaccination has reduced the cases of human salmonella by

half. [1] EU changes to the Consumer Protection Act make primary food producers liable, so it is a good idea for all egg and table bird producers to have public liability insurance. [2]

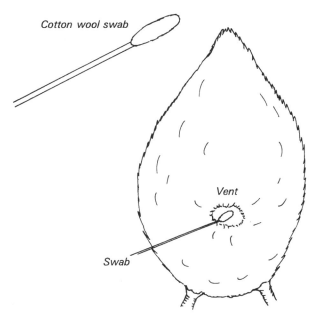

Source: Greendale Laboratories Ltd.

Figure 13.1 Method of taking a swab.

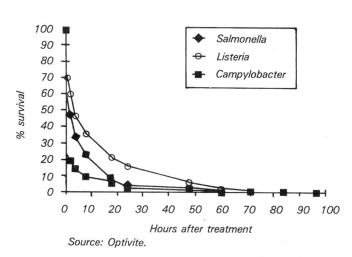

Source: Optivite.

Figure 13.2 Effect of organic acid additives on potential contamination of animal feeds.

Poultry zoonoses

It is important to remember that there are some organisms which can be transferred from poultry to people.

Newcastle disease can affect humans, causing conjunctivitis and mild flu-like symptoms.

Cryptosporidium is an organism that can be transmitted from poultry or their housing to the hands, and then transferred to the mouth. It causes diarrhoea so the hands should always be washed after handling poultry or managing their houses.

Organisms causing salmonella, campylobacter and listeria can also be transferred to humans.

Although not directly related to poultry, Weil's disease from rats which are attracted to the site can be passed to man. It is transmitted in the rat urine which may contaminate surfaces that are subsequently handled.

Aspergillosis, graphically called 'farmer's lung', is caused by a fungus in damp hay or feed. Hay should never be used as a nest box liner.

Notifiable diseases

The following are notifiable diseases and if they occur, they must be reported to the DEFRA: Newcastle disease and avian influenza. Paramyxovirus in pigeons, also a notifiable condition, not only affects feral, racing and show pigeons, but can also cause Newcastle disease in poultry.

An A–Z of poultry diseases and problems

Aspergillosis (Fungal pneumonia, Pulmonary mycosis, Farmer's lung)

Caused by the fungus *Aspergillus fumigatus*, this condition is brought about by inhalation of the spores from contaminated litter or feed. Symptoms are excessive thirst, gasping and rapid breathing, with an overall depressed posture. Young birds are particularly at risk, and there is no effective treatment, although antibiotics have been shown to produce an improvement. Strict hygiene, avoidance of damp hay, straw, wood shavings and feed, together with good management of litter, are needed. The condition can also affect man.

Avian encephalomyelitis (AE, Epidemic tremor)

This condition is caused by a virus which is transmitted primarily through the egg. The only effective control is by vaccination of breeders. The disease is seen mainly in chicks from 1 to 3 weeks old. Movements are restricted and trembling of the head and neck can be seen.

Breeding flocks which are affected will acquire an immunity, but no eggs should be incubated from them for several weeks until this is established.

Avian influenza

Avian influenza is caused by airborne viruses and affects the respiratory tract, as does the common cold in man. There may be a slight swelling of the head and neck, and a nasal discharge is usually seen. Mortality is normally low with mild forms, but there are more severe strains. There is no treatment, although antibiotics have been used where there are secondary infections.

Blackhead (Enterohepatitis, Histomoniasis)

More usually associated with turkeys, blackhead can also affect chickens. The causative agent is a protozoan parasite called *Histomonas meleagridis* which is transmitted via water, feed or droppings. Infected eggs of the parasitic caecal worm, *Heterakis gallinarum*, are also a source of infection. It is important not to allow ground to become over-used, or to have chickens and turkeys sharing the same land.

Bumblefoot (Abscess of the foot)

Bumblefoot is the common name for the swelling which results from an infected cut or graze on the underside of the foot. The wound heals on the outside, leaving a hard core of pus on the inside. It is sometimes found where birds are provided with perches which are too high for them, or where the grazing area is on flinty ground. The affected bird will have a limp, and examination of the foot reveals the hard abscess. Applying slight pressure is sometimes enough to burst it, releasing the pus, but it may require lancing with a sterilised blade. Antiseptic liquid or hydrogen peroxide should then be applied to the affected area. The latter is very effective in oxygenating the wound as well as bubbling out the dirt.

Caecal worms (Heterakis)

Caecal worms are about 1.5 cm long and inhabit the caeca. They do not cause disease, although their eggs are capable of transmitting blackhead. They can be controlled by means of the poultry wormer Flubenvet.

Chronic respiratory disease (CRD, Airsacculitis)

Sneezing, coughing and wheezing are the signs which indicate this condition. It is initially caused by viral infection, followed by secondary bacterial invasion of the organism *Mycoplasma gallisepticum*. The best procedure is avoidance, by ensuring adequate ventilation in the house and a stress-free management system. Mild cases, where secondary infection is slight, will clear up fairly quickly, but severe cases may require antibiotic treatment.

As the tendency can be inherited via the egg, it is important to ensure that all breeding stock has been blood tested and found to be healthy. The symptoms are similar to those of Newcastle disease and infectious bronchitis, so if in doubt, consult a veterinary practitioner.

Cloactitis (Vent gleet, Poultry venereal disease)

The signs of this contagious venereal disease are swollen membranes in the cloaca, with a whitish discharge. The affected bird should be separated from the flock and the vent area painted with iodine. The bird may recover; otherwise it should be culled. The disease is most commonly found in hens which are allowed to run indiscriminately with an infected male. The sexes should be kept quite separate unless controlled breeding is required, and the breeding stock has been checked for health. Cloactitis may occasionally be found in a hen that has been subject to vent pecking.

Coccidiosis

Coccidiosis is caused by protozoan parasites which are transmitted via infected droppings. Affected birds are listless, have pale combs and pass blood in the droppings. Mortality can be as high as 50 per cent. Coccidiocidal agents are available where there is an outbreak, but every effort should be made to change grazing areas frequently, and to deal with damp areas of litter.

Coryza (Infectious coryza, Roup)

Caused by the bacterium *Haemophilus gallinarum*, this condition is best avoided by scrupulous attention to good ventilation in the house and general cleanliness in relation to equipment and environment.

Symptoms are similar to chronic respiratory disease, with inflammation of the eyes and nose, and sneezing and wheezing. Severe cases may require antibiotic treatment but mortality is generally low.

Egg drop syndrome (EDS)

The causative virus for this disease is transmitted through the egg. If carrier birds are subsequently incorporated into a flock, their droppings may provide a source of infection for the rest of the flock. Egg production is affected, with some deformed

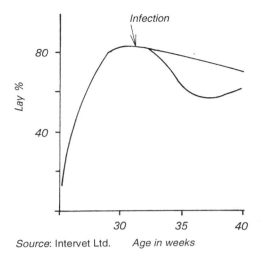

Figure 13.3   Egg drop by infection during lay.

eggs being produced. The effect on egg production is shown in Figure 13.3.

There is no treatment, but a vaccination at the point of lay period provides protection.

Egg peritonitis

When an egg leaves the ovary it is caught by the opening of the egg canal, called the infundibulum. If it misses and slips into the surrounding peritoneal cavity, it becomes lodged, goes bad and causes peritonitis. There is nothing that can be done, and the hen should be put down humanely.

Fowl cholera (*Pasteurellosis*)

Caused by the bacterium *Pasteurella multocida*, fowl cholera is recognised by swollen, bluish wattles in severely affected birds. It does not cause high mortality and can be treated with antibiotics in the drinking water. It is often carried by vermin so efforts should be made to deter them from the site.

Fowl pox (Avian pox, Avian diphtheria)

This virus infection is transmitted by bird contact and mosquitoes or via water and food. There are characteristic lesions (pox marks) on the comb and there may also be laboured breathing. Mortality is usually low if the only symptoms are the lesions, but the more serious form which is accompanied by lung congestion causes around 40 per cent mortality. If the disease is detected, it is advisable to vaccinate the whole flock without delay.

Fowl typhoid (Pullorum disease)

This is caused by the bacterium *Salmonella pullorum*, and affected birds produce white diarrhoea. Unless treated with antibiotics it can cause quite high mortality levels in the flock. It can also be transmitted via the egg, hence the importance of breeding only from blood tested, healthy birds.

Gumboro disease (Infectious bursal disease, IBD)

This is a highly infectious viral disease which is difficult to eradicate from a site. There is no treatment, but vaccination will give protection against it.

Hairworms (Capillaria)

These internal worms cause damage to the lining of the intestine, resulting in anaemia and pale egg yolks. A suitable poultry wormer such as Flubenvet can be given.

Infectious anaemia (Inclusion body hepatitis)

Transmitted via the egg and droppings, this virus causes listlessness and some mortalities in the initial stages. If recovery takes place, the birds should not be allowed to breed. There is no treatment, although antibiotics can be used to control secondary infections.

Infectious bronchitis (IB)

This is a viral infection for which there is no treatment and chicks are normally vaccinated at an early stage. Where unvaccinated birds contract the disease any eggs laid will be deformed and, occasionally, devoid of shells. Birds that are vaccinated will occasionally contract different strains and a sudden drop in egg size may be an indication.

Infectious laryngotracheitis (ILT)

Caused by a virus of the herpes group, ILT does not normally cause mortalities, but egg production in laying flocks may drop for about three weeks. Wheeziness and laboured breathing are signs of infection. A vaccine is available to provide protection against it, but if scrupulous attention is paid to general hygiene, with adequate cleaning of feeders and drinkers, there is little chance of birds encountering it.

Lice and mites

These can be troublesome, particularly in the summer. Red mite comes out of its hiding places at night under perches and in the woodwork, to feed on the birds. Mites are also responsible for scaly leg. Treat with a proprietary product such as Ban-Mite from a licensed supplier. (See the section on summer management on page 98.)

Lymphoid leucosis (LL, Big liver disease, Visceral leucosis)

This is a disease which is thought to be transmitted via the egg. It is essential to breed only from birds which are found to be clear of it, following a blood test. Affected birds develop large tumours, sometimes with enlarging of the bones and wings. There is no cure.

Malabsorption syndrome (Pale bird syndrome, Stunting disease)

Little is known about this disease, but it is more frequently seen in the broiler industry than in the free-range sector. Chicks develop ricket-like symptoms, and have extremely pale heads and legs. There is also diarrhoea with brown, foamy droppings. There is currently no treatment. Hygienic conditions and fresh air with access to sunshine should prevent its appearance in outdoor birds.

Marek's disease (MD, Neurolymphomatosis)

This is another herpes virus disease which is transmitted through the egg. Again, all breeders should be blood tested to ensure that they are free of it before being allowed to breed. Paralysis and tumours develop before death occurs. Day-old chicks are normally vaccinated against it.

Newcastle disease (ND, Fowl pest)

This is a highly infectious disease caused by a paromyxovirus, and poultry is normally vaccinated against it for there is no treatment. Clinical signs are wheezing and choking and the onset of paralysis. It is a notifiable disease, so the Ministry of Agriculture must be informed. Where an outbreak occurs there is a policy of flock slaughter.

Roundworms (Ascarids)

Found in the gut, these parasitic worms are around 5 cm in length, and although a healthy bird can tolerate a certain burden, if there are too many the result is anaemia and pale yolks. A poultry wormer such as Flubenvet is effective in getting rid of them.

Tapeworms

Tapeworms grow to around 10 cm in the gut, but cause comparatively little harm unless the burden becomes too great. In this case, the bird needs to eat more than it would normally in order to maintain its metabolism. Flubenvet from the vet or licensed suppliers can be administered in severe cases.

Commercial producers will be liaising with their vets on a regular basis, but the small flock owner is in a different situation. In some areas, vets are not particularly experienced with poultry, and it is obviously not always practical or cost-effective to take a single bird for a consultation. Recommended books for advice and information for small poultry keepers are *Diseases of Free Range Poultry* by Victoria Roberts (UK) and *The Chicken Health Handbook* by Gail Damerow (USA).

References

1. Advisory Committee on Microbiological Safety of Food report, 2001.
2. EU regulation update to Consumer Protection Act, 2000.

Appendices

KEEPING RECORDS

Keeping adequate records is essential in any activity. A free-range enterprise is no exception, and indeed there is a legal requirement for the commercial producer to keep careful records.

ADAS and several of the feed companies run costings services for the large producer. These schemes enable both performance and cash flow records to be monitored accurately, as well as providing advisory facilities on management programmes.

There are also computer packages available for those who prefer to use a PC, including one called Egg Stat which is specifically for free-range egg production. Alternatively, any database can be used for a wide range of records. Many agricultural colleges and local authorities run short courses on learning to use a computer.

Even the smallest poultry keeper who is not necessarily selling eggs or table birds would be wise to keep records. There are also records for breeding and other activities associated with keeping and showing poultry.

Egg records

If a producer is registered to sell eggs described as free-range, there is a legal requirement to keep the following records:

- the date of placing hens in the house
- the age of hens at placing

- the number of birds by type of poultry system
- the number of eggs produced and delivered each day and the date of despatch of the eggs to the collector or packing station, with the name, address and packing station number of the purchasers

If a producer of free-range eggs is registered to grade and pack his own eggs, in addition to the records above, he is required to keep records of the following:

- daily quality and weight grading
- sales of eggs and small packs
- the number of packs
- type, number and/or weight of eggs sold by weight grade
- date of despatch and delivery
- stock records on a weekly basis

Comprehensive details are available from the Regional Egg Marketing Inspector.

There are many ways of recording details. One example is shown on page 157.

On a small scale, it helps to have a calendar hanging up in a convenient place, so that the total of eggs (and breakages) can be jotted down immediately. Numbers can be transferred to a more permanent record book later. It is easy to think that the totals will be remembered, but they never are. Unless they are written down every day, they will be forgotten!

FARMGATE EGG RECORD

House No. _____ No. of birds _____ Breed _____

Date Housed _____ Age Housed _____

DATE OF MONTH	MONTH						MONTH					
	EGG COLLECTIONS			DAILY TOTAL EGGS	FEED		EGG COLLECTIONS			DAILY TOTAL EGGS	FEED	
	1	2	3		Qty. Fed/Day kg/lb	Mortality & Culls	1	2	3		Qty. Fed/Day kg/lb	Mortality & Culls
1												
2												
3												
4												
5												
6												
25												
26												
27												
28												
29												
30												
31												
TOTAL												

NOTE: 1 kg = 2.2 lb; 1 tonne = 1000 kg = 2205 lb. 1 Bag FARMGATE LAYERS = 25 kg = 55 lb approx.

BOCM SILCOCK

Egg production record.

MONTHLY TRAP NEST RECORD

Month _____ Breed _____ Ring No _____

Weekly Total

Monthly Total

Monthly trap-nest record.

Incubator Record

Date set _____   Date due _____

Breed _____

Parent Breeders _____

Leg Ring Numbers _____ _____

No of eggs set _____   No of infertiles _____

No of chicks hatched _____

No of chicks dead in shell _____

No of chicks removed to brooder _____

Notes:

Incubator record.

Table bird records

The first thing to record is the number and source of birds in the batch. This may be a group bought in as day-old chicks, hatched on your own premises or bought in as six-week-old pullets. If they are to be raised organically, they must be day-olds. Any mortalities should be recorded so that the final production statistics relate to the original number of birds. Record the date at which batch production commences and, thereafter, the amount of feed consumed every day, or at whatever period is appropriate for your own management system. Again, it is necessary to record details of feeds if the birds are to be sold.

My own practice, where birds were raised for the family freezer, was to record every day in a notebook hanging up in the feedstore, and to transfer data once a week to a more permanent record book.

It may be necessary to keep regular checks on weight so that the amount of weight gain in relation to feed consumed can be calculated. Large broiler units often use electronic weighing systems which will then work out the feed conversion ratio – the average amount of feed consumed in relation to average weight gain. On a small scale, weighing a few birds from each batch, once a week, is usually sufficient. A suspended spring balance with a canvas bag is effective. The canvas keeps the wings confined, while the hole at the bottom allows the bird's head to emerge, keeping stress to a minimum.

The feed conversion ratio can be worked out by seeing how much food is eaten for every 450 g (1 lb) weight gain. If, for example, 1.35 kg (3 lb) of food is eaten for every 450 g (1 lb) of weight achieved, the feed conversion would be as follows: FCR = 3:1. The average intensive broiler unit aims for a feed conversion of around 2, with birds reaching a weight of approximately 2.05 kg (4.5 lb) in 45 days. Obviously, such an aim is meaningless to a free-range enterprise where slow growing birds are kept for about 12 weeks, but the general principle of recording grain consumption in relation to weight gain in order to find the feed conversion ratio is just as valid.

A simple recording system which I used for my own table birds is given below.

Date commenced		Comments		
Number in batch				
	Feed consumed	Average weight gain	Mortalities	FCR
Week 1				
Week 2				
Week 3				
etc.				
Note: It is obviously necessary to begin by recording the average weight of the batch if they have been bought in as pullets.				

A simple recording system for table birds.

Feed records

On a commercial scale, it is necessary to record the nutrient content of feed, as declared by the feed compounder. Organic producers will also need to keep records of any nutrients that are bought in from both organic and non-organic sources.

On a small scale, it is easy to record every time a sack of feed is opened, as long as there is a convenient hook on the door of the feedstore. When the sack is opened, tear off the label and put it on the hook. When you come to writing down the information in a record book, it is then just a matter of looking on the hook. Be sure that the price of the sack is recorded as well.

Larger producers may find it more difficult to record the amount of feed used in a given time, particularly if bulk deliveries and an automatic feed system are used. The solution here is to install a feed weigher so that regular checks of feed provision can be made. Feed intake can then be checked manually once a week by hand-filling suspended feeders or chain feed hoppers and seeing how much is eaten in a given time. If there is a meter for feed bin readings, this will not be necessary.

Record of veterinary medicines

On a commercial scale, any veterinary medicine used will need to be recorded, together with the date and details of when it was administered. If there is a stated withdrawal period, this must be recorded and observed. A withdrawal period is the time that must elapse between the administration of the last dose of medicine and the sale of the eggs or meat.

There are two categories of medicines. Pharmacy and Merchants List (PML) products are available without prescription, but only from merchants who are registered to sell them. Examples are external sprays or powder for treating against lice or mites. Prescription Only Medicines (POM) are available only on prescription from a veterinary surgeon. These include products such as antibiotics.

The domestic poultry keeper is also well advised to keep records of any medicines that are used, although there is no requirement to do so.

Financial records

It goes without saying that for a commercial enterprise, all purchases, transactions and sales should be carefully recorded. The way in which this is done is best organised after consultation with an accountant. Any business should have the services of a good accountant from the start. Apart from drawing up and presenting your accounts to the Inland Revenue, he is in a good position to offer sound advice on how to save money and possibly avoid some tax. (Avoiding tax is a legitimate activity, not to be confused with evading tax, which is unlawful.)

Bear in mind that it is necessary to be registered for VAT if your turnover is likely to be in excess of the statutory amount for that year. There is currently no VAT on foodstuffs so you will not be charging this tax on eggs or table poultry, but if registered you will be able to claim back all VAT on items you purchase that are relevant to the business. For example, stationery, packaging and services all carry VAT. Keep a careful record of these and claim on your VAT return. Those who use computers can use software such as QuickBooks to keep track of all their financial records.

Insurance cover is also essential in any business. Find a reliable and registered insurance broker and discuss with him the whole nature of your enterprise at an early stage. He will be able to suggest all sorts of aspects which you may not have considered. Those who have farm shops, where the public have access, for example, will need a comprehensive public liability coverage against accidents and other eventualities.

There are specialist companies which offer cover on livestock and farming activities.

Special Marketing Terms (SMTs)

There are four descriptions used for marketing eggs that are produced in non-battery systems:

Free-range   maximum 1,000 birds per hectare
Semi-intensive   4,0000 birds per hectare
Deep litter   7 birds per square metre
Perchery (barn) 25 birds per square metre

In areas of the country which are particularly cold in winter, some producers keep their birds inside until weather conditions improve. Eggs from birds that are confined in this way cannot be described as free-range, so an alternative must be used. Some producers have two or more houses, producing free-range and perchery/barn eggs. Needless to say, great care needs to be exercised in this case, so that the eggs are kept separate. It is not unknown for prosecutions to take place where eggs sold as free-range have turned out to be from an alternative system.

CHICKEN FACTS AND FIGURES

The average hen

- has a temperature of 39°C
- operates at an optimum temperature of 21°C
- eats around 130–150 gm a day
- drinks approximately 20 ml of water a day (more in hot weather)
- incubates eggs for 21 days
- looks after her chicks for up to 5 weeks

How long does she live?

How long a hen lives is anyone's guess, but the following experiences may shed some light on the matter:

'My oldest hen is 9 years old now, still sprightly and still laying.'
Source: Dr C.A. Keepax writing on the Scots Dumpy, Home Farm, March 1992.

'I have a brown hybrid hen who is 5½ years old. She has consistently laid 5–6 eggs a week in the laying period and the only difference is that recently the eggs have become more fragile.'
Source: Sue Malone, Home Farm, November 1993.

'I have amongst my flock, two of my original Cuckoo Maran bantams, both coming up to 9 years old.'
Source: Celia Surrell, Home Farm, January 1994.

'I have a hen, Tiny, who will be 12 this summer. She is a small hen, although not truly a bantam'.
Source: Patsy Therleigh, Country Garden & Smallholding, June 1994.

'Sadly, yesterday, we buried our old Irene, a cross bantam. She was 15 years old.'
Source: Mr & Mrs Hughes, Country Garden & Smallholding, July 1994.

A tale of survival

Poultry producers may think that chickens have a tendency to expire, but they can also be extremely resilient, as the following story illustrates:

An Ancona hen called Paxo disappeared for six days and was eventually found wedged in between some hay bales. The temperature had been baking hot. When she was discovered, she was light and could not walk, but after a few days in a small coop with food *and drink, she was fine and was able to rejoin her friend Hetty.*

Source: Claire Greenhow, *Home Farm*, Issue 49, 1983.

Poisonous to poultry

Chickens are fairly discriminating about what they eat, and they are unlikely to harm themselves by eating poisonous plants. Nevertheless, it pays to avoid risks where free-ranging birds are concerned. The following are known to be poisonous to poultry.

Bryony, *Bryonia cretica*
Deadly nightshade, *Atropa belladonna*
Horseradish, *Armoracia rusticana*
Hemlock, *Conium maculatum*
Henbane, *Hyoscyamus niger*
Laburnum, *Laburnum anagyroides*
Monkshood, *Aconitum napellus*
Privet, *Ligustrum vulgare*
Rhubarb, *Rheum palmatum*
Yew, *Taxus baccata*

Traditional sayings

There is a wealth of traditional sayings with allusions to poultry, although as society has become more urbanised, they have become less common.

Don't count your chickens before they are hatched.

Never crack your eggs until you have your salt.

As bare as an egg.

It is very hard to share an egg.

Better half an egg than an empty shell.

Sure as eggs are eggs.

You may get a black chick from a white egg.

They were hatched in the same nest.

Every bird must hatch its own egg. (Fuller)

Don't put all your eggs in one basket.

Every time the cock crows, he loses a mouthful. (Fuller)

Your last chick is hatched.

A whistling woman and a crowing hen, are neither good to God nor men.

He that will steal an egg will steal an ox.

Teach your grandmother to suck eggs.

Let fowl go afield, and other things taste. (Tusser)

It is an ill hen that will not earn its own keep.

He is like a cock, always crowing.

That one always has to be cock of the walk.

Feed a chick and you'll have a hen.

He eats the chick in the shell.

He blushes like a black hen.

I would not have your cacklings for your eggs.

'Mine honest friend, will you take eggs for money?' (Winter's Tale, Shakespeare)

Higgledy, piggledy, my black hen,
She lays eggs for gentlemen;
Sometimes eight, and sometimes ten,
Higgledy, piggledy, my black hen.
(Traditional nursery rhyme)

A mine of information, miscellaneous facts and popular beliefs about chickens can be found in *Poultry of the World* by Loyl Stromberg.

GLOSSARY

This section provides definitions of most of the terms likely to be encountered in the world of poultry. Please note that terms relating to health and disease are in Chapter 14.

Abdomen   Area under the body, from breast to stern.

Addled eggs   Fertile eggs that die soon after the commencement of incubation.

Ad-lib feeding   Making food available so that birds can feed when they want to.

Air cell   The area containing oxygen at the broad end of the egg.

Albumen   The white protein area of the egg.

Alimentary tract   The specialised tube, from mouth to vent, where the digestion, assimilation and egestion of food take place.

Amino acids   Digested proteins ready for assimilation into the bloodstream.

Antibiotics   Internal medication for combatting bacterial infections.

Antibodies   Microscopic agents formed in the bone marrow and spleen and circulated in the blood as defence mechanisms against viral disease.

AOC   Any, or all other colours, as in a breed.

AOV   Any, or all other varieties, as in a breed.

Ark   Moveable range unit with a slatted floor, traditionally used for growers.

Artificial insemination (AI)   The collection of semen in a glass tube from the male, followed by its injection via a syringe into the vent and oviduct of the female.

As hatched (AH)   Where no attempt has been made to sex chicks.

Ash   The mineral component of feed.

Autosexing breed   One that produces readily identifiable sexes at hatching because of the barring on the plumage.

Axial feather   Small feather between primary and secondary feathers of the wing.

Baffle   Board used in a house, in conjunction with an air vent, to prevent down-draught without restricting ventilation.

Bantams   Small sized fowl, either true bantams or miniaturised versions of large fowl.

Barn eggs   (See Perchery/barn eggs.)

Barring   Where stripes of another colour run across the feathers, as in Plymouth Rock and Scots Grey. Openly-barred is far-apart stripes of a different colour. Also refers to the sex-linked barring gene.

Battery system   An intensive system of egg production where layers are kept in cages in environmentally controlled houses.

Beak   Horny projection consisting of upper and lower mandibles, forming the mouth parts.

Beak trimming   Removal of the tip of the upper beak to stop pecking and cannibalism. Not recommended on humanitarian grounds.

Beard   A tuft of feathers under the beak of some breeds, such as the Creve-Coeur.

Bird   General term referring to any individual or breed.

Blade   The lower, undivided part of a single comb.

Booted   Feathers on the shanks and toes, or having vulture hocks.

Breast   Area of the body from the neck to the central part of the body between the legs. Also the meat on either side of the keel, in a table chicken.

Breed   A group of birds with similar appearance and characteristics, which produce offspring identical to themselves when mated together. A breed may include different varieties, with varying colours, combs, etc.

Broiler   Fowl bred and raised for the table.

Brooder   A protected area with artificial heat for raising chicks.

Broodiness   The behavioural pattern in females of wishing to sit on, incubate and brood a clutch of eggs.

Candling   Shining a bright light through an egg in order to see the internal contents.

Cape   Feathers between the shoulders and under the hackles.

Capon   (See Caponising.)

Caponising   The practice of castrating young cockerels, surgically or chemically, to produce capons for the table. The practice is now illegal in the EU.

Carbohydrates   Heat and energy producing foods found in grains and vegetables. Soluble carbohydrates are sugar and starch. Fibrous carbohydrates refer to crude fibre. (See Fibre.)

Carriage The way in which a bird holds itself and walks. An important aspect of show requirements, as well as in selection of breeders.

Chalazae Internal membranes in the white of egg, that keep the yolk suspended in the centre.

Chaff The outside husks of wheat and oats.

Chick Young bird, of either sex, up to the age of 6 weeks.

Chick crumbs Proprietary feed or starter ration formulated for chicks.

Chicken Originally a term for a male or female up to the age of 12 months. Now a general term for all ages. Also a description of the meat from a table bird.

Clavicle Wishbone.

Clears Eggs that are found to be infertile on candling, and therefore removed from the incubator.

Cloaca The opening at the end of the rectum.

Close feathering Where feathers are held close to the body, but not the same as hard feathering, as in game birds.

Clutch The number of eggs laid by one hen.

Cock Male over the age of 12 months.

Cockerel Male bird up to the age of one year.

Cock-fighting Sport using game birds to fight each other. Illegal in many parts of the world, it still continues in parts of South America and Europe.

Colony house Moveable house on wheels for a small flock.

Comb Fleshy growth on top of the head, seen in different forms and either red or black.

Condition The general state of a bird in relation to health, cleanliness and appearance.

Coop Small house, with or without run, for a hen and chicks, or other purpose such as temporary accommodation.

Coverts Tail coverts are the small feathers at the base of the tail. Wing coverts are those covering the tops of the flight feathers.

Crest Tuft of feathers on the top of the head, as found in Creve-Coeur, Faverolle, Houdan and Poland.

Crop Area of the lower gullet where food is stored before passing into the gizzard for further digestion.

Cross breeding The mating of two birds of different breeds, e.g. Rhode Island Red x Light Sussex.

Crow-headed Narrow, shallow head and beak.

A serious defect in show birds, also an indicator of a poor layer.

Cuckoo Feather markings similar to barring.

Culling Disposing of birds.

Cushion The soft feathers on the hen's rump. (See Saddle.)

Custom hatching Incubating birds to customer order.

Cuticle The protective bloom left on a new laid egg as it dries.

Dam Female parent.

Day-olds Newly hatched chicks. They can survive without food for 48 hours because of the remnants of the yolk in the abdomen. Transporting day-olds is therefore straightforward.

Dead-in-shell Chick embryos which began to develop in the shell but died before hatching.

Deep litter A system of housing layers on litter such as wood shavings, chopped straw, etc, where flock density does not exceed 7 hens per square metre.

Defect Any aspect that detracts from perfection in the standards of show birds, or a genetic fault in general terms.

Dominant Genetic characteristic that appears in the progeny of a first hybridised cross, e.g. silver is dominant to gold.

Down Soft feather covering of the newly hatched chick. Also under-feathering with no hooks.

Drawing Eviscerating or gutting a chicken.

Dressing Preparing a bird for the oven.

Droppings board (or pit) A sliding board placed under the perch to catch droppings. A variation on the principle is to have a droppings pit under the perch, or under a slatted floor in a large house. The board can be removed periodically to remove the droppings.

Drumstick Tibia or lower thigh, as referred to in a meat bird.

Dual-purpose breed A utility bird that is a good layer and is also suitable for the table.

Dubbing Traditional practice of removing the cock's comb (and sometimes wattles and earlobes in fighting birds) where over-large and interfering with feeding. No longer allowed on humanitarian and legal grounds in Britain. If frostbite has damaged and blackened the comb, veterinary advice should be sought.

Dustbath An area of fine soil which the bird allows to trickle through the feathers. Dust bathing is instinctive behaviour in attempting to get rid of external parasites.

Ear lobes   Fleshy areas below the ears. Where they are white, the eggs produced by the bird are always white.

Ectoparasites   External parasites such as mites and ticks.

Egg box   Compressed fibre or plastic containers, with lids, for 6 or 12 eggs.

Egg case   Container of 360 eggs.

Egg grading   A system of quality checks for eggs being sold for human consumption.

Egg sanitant   An antiseptic solution for ensuring that hatching eggs are free of external pathogens.

Egg sizes   Sizes that are recognised for the purposes of labelling and selling.

Egg-tooth   (See Pipping.)

Egg tray   (See Keyes tray.)

Endoparasites   Internal parasites such as worms and coccidia.

Face   The area around and below the eyes where no feathers grow.

Fancier   One who breeds birds for exhibition.

Feathers   Outgrowths from the skin epidermis which collectively form the plumage.

Wing showing primary and secondary feathers.

Feather-legged   Having feathers on the shanks and toes.

Feed Conversion Ratio (FCR)   Ratio of food eaten in relation to liveweight or number of eggs produced.

Fibre (Fib)   The indigestible element of food, provided by cellulose and lignin in grain, that enables the bowels to work efficiently.

First-cross   The progeny of two different breeds mated together.

Flights   The long feathers of the wings (also called primaries.)

Flock mating   Crossing a number of males with a flock in the same pen, as a means of producing a large number of fertile eggs.

Fluff   Soft feathering on the abdomen and thighs, more developed in heavy than in light breeds.

Fold unit   Moveable house on grass, often used for young birds.

Fount (or fountain) drinker   Two part container, one a shallow dish from which birds drink, the other a reservoir which automatically tops it up.

Fowl   Originally chickens of all ages and sex, now also used as a general name for domestic chickens, ducks, geese and turkeys.

Franchise   An operation whereby a producer is supplied with housing, equipment and stock for a certain investment, and has the produce collected and distributed on his behalf.

Free-range   A system of keeping birds with access to grass at a maximum density of 1,000 birds per hectare.

Frizzle   Feathers which curve forwards and outwards. Also the name of a breed.

Gall bladder   Small organ producing secretions for the digestive system. It is important to remove it when eviscerating a table bird; otherwise it adds a bitter taint to the meat.

Gapes   The action of a bird in opening and closing its beak in a gasping action, when it is affected by parasitic nematodes in the windpipe.

Genotype   The genetic makeup of a bird. (See also Phenotype.)

Gizzard   The true stomach of a bird, with strong, muscular walls to assist in the breaking down of grains.

Grit   Solid mineral particles that help digestion in the gizzard.

Grower   Young bird between 6 weeks and maturity.

Grower ration   Compound feed formulated for the needs of growing birds.

Hackles   Long, narrow neck feathers. Tail hackles are the narrow ones on the saddle of a cock.

Handling   The art of picking up and holding a bird without causing stress.

Hangers   Thin feathers that hang down at the root of the male's tail.

Hard feathering   Tight, close feathering, looking as if plumage has been brushed and then stuck down. A feature of game birds. (See also Close feathering.)

Hardening off   The period when young birds no longer have artificial heat, but are not yet ready to go outside.

Hatching   Emergence of the chick from its shell.

Haugh unit   Measurement of the height of the egg white, as used to denote freshness.

Heavy breed   Large breed or sitter with a greater tendency to become broody.

Hen   Female bird over 12 months.

Hen-housed average   Number of eggs laid by a flock, divided by the number of birds when first housed, including mortalities.

Hock   The joint between the lower thigh and shank, often mistakenly referred to as the knee joint.

Hopper   Feeder made of two parts, a container from which birds can feed, and a reservoir which automatically tops it up.

Hover   A heated roof or canopy for brooding chicks. (See Brooder.)

Hybrid   Progeny produced by breeding from two or more distinct lines.

In-breeding   Mating birds that are brothers or sisters.

Incubation   The process of development of an embryo in the shell for 21 days.

Joule   Unit of measurement of energy levels in feeds. (See also Metabolisable energy.)

Keel   The breast bone to which the flying muscles of the wings are attached.

Keyes tray   A fibre or plastic tray holding 30 eggs.

Kibbling   Chopping up grain into smaller particles, as distinct from grinding it.

Knee   Joint between the upper and lower thighs, not to be confused with the hock.

Label Rouge (Red Label)   A marketing description to identify French table birds grown on free-range.

Lacing   Pattern on the plumage where the outside edge of a feather is of a different colour or shade from the rest. It may be single or double.

Laying trials   Competitions that indicate the relative productive capabilities of different breeds. The first one was held in 1897 in Britain.

Leader   Tapering spike or end point at the rear of a rose-comb.

Leaf comb   A comb resembling a broad leaf, as in the Houdan.

Leg   Area consisting of upper and lower thigh and shank.

Leg feathering   Where feathers are found instead of the usual scales, as in the Cochin, Booted Bantam and Maline.

Light breed   Lighter in weight than a heavy breed, and with less of a tendency to become broody.

Litter   General term for ground covering, e.g., wood shavings, straw.

Livability   The ability to survive to a given age.

Maize   A grain constituent of poultry feeds, used in various forms, such as gluten and meal. Also called sweetcorn in the UK and corn in the USA.

Mandible   The lower, moveable part of the beak.

Marbling   Spotted pattern on the feathers.

Mash   Compound feed made of grains and other food materials.

Mealy-feathered   A fault in a show bird, where brown or buff feathers are spotted with white.

Metabolisable energy (ME)   The amount of energy provided by food, quoted by megajoules/kg. (See also Joule.)

Middlings (See Wheat.)

Mossy-feathered   White feathers spotted with brown. A fault in a show bird.

Moult   The annual process of losing old feathers and growing new ones.

Mottling   White tipping on the ends of feathers, as in the Ancona. (See also Tipping.)

Muff   Feathering on each side of the face, as in the Faverolle.

Natural brooding   Where chicks are protected by a mother hen, rather than being raised in an artificial brooder.

Nostrils   Openings at the base of the upper beak for respiration.

Notifiable disease   One whose presence must be notified to the Ministry of Agriculture, Fisheries & Food, or to the police.

Oats   Grain used in feeds. Too much causes over-heating and feather loss.

Oesophagus   Gullet or area of the digestive system from the mouth to the crop.

Out-breeding   Mating different strains of the same breed.

Oviduct   Passage from the ovary to the vent, down which the egg travels.

Oystershell   When crushed to a suitable size, provides a good source of calcium.

Packer (Packing station)   Site registered with the Regional Egg Marketing Board to grade and pack eggs.

Parson's nose   The uropygium, a bony and fleshy protuberance from which the tail feathers grow.

Pathogen   An organism that causes disease.

Pea comb   A small, low triple comb, as in the Brahma.

Pecking order   Hierarchial scale within the flock.

Pellets Compound mash feed compressed into pellets.

Pencilling Line markings on the feathers.

Pen mating Crossing one male with a number of females in a pen.

Perchery (barn) eggs System of housing layers where flock density does not exceed 25 birds per square metre, with a minimum of 15 cm perch space per bird.

Phenotype The physical type, as seen outwardly. (See also Genotype.)

Pipping Where a chick breaks a hole in the egg shell, prior to hatching, with a horny tip to its beak (sometimes called an egg-tooth).

Pile A term to describe a breed or variety where the male is piebald and the hen of one colour, or two shaded into each other, as in Pile Game and Pile Leghorn.

Pituitary gland Small gland in the brain which controls the egg-laying mechanism by responding to day length.

Plumage (See Feathers.)

Point-of-lay (POL) The period from 18 to 21 weeks, when a pullet is about to lay for the first time.

Pollards Middlings (See Wheat.)

Pop-hole A poultry house exit that can be opened or closed for the birds as required.

Pot egg Pottery egg placed in a nest box to encourage hens to lay.

Poussin (or Petit poussin) A table bird killed at an early stage for a specialist market.

Precocity Starting to lay before growth is complete.

Primaries (See Flights.)

Progeny testing A way of assessing breeders by checking the performance of their offspring. Egg size, for example, is passed through the male line.

Proventriculus The glandular stomach area before the gizzard.

Pullet Originally a show term for a female exhibited in the first year of hatch, it is now generally taken to be a female from the age of 6 weeks to the commencement of laying.

Pure breed One that will breed true so that the progeny resemble the parents. (See Breed.)

Range shelter A lightweight structure with mesh walls, roof, door and perches, traditionally used for growing stock. Simplified versions of roof and pole walls only are now used to provide shade and shelter for free-ranging flocks at a distance from their house.

Recessive A gene that is masked by a more dominant one in a first hybrid generation.

Rectum The last section of the alimentary tract.

Replacements New stock to replace older birds.

Reversion Tendency to go back to an original trait.

Roach back A humped back condition which is hereditary. Avoid breeding from birds with this trait.

Roche scale Measurement of degree of yellowness in yolk.

Rooster Old English and American term for a cock.

Saddle Back of the male. (See also Cushion.)

Secondaries Inner quill feather of the wings, second in size to the primaries.

Seconds Middlings (See Wheat.)

Self colour Single plumage colour throughout, such as white, black or buff.

Semi-intensive System of keeping layers where flock density is no greater than 4,000 birds per hectare.

Sex-link A cross between two different breeds where the chicks can be sexually identified.

Scales Thin slivers of horny tissue covering the legs.

Serrations Divisions of the comb.

Setting Fertile eggs that are being stored while awaiting incubation.

Shank Area of the leg between the foot and the hock joint.

Sharps (See Wheat.)

Sheen Gloss on the plumage.

Sire Male parent.

Sibs The offspring of a brother-sister mating.

Sickles Long, curved tail feathers.

Single comb Narrow, thin comb surmounted by serrations.

Sitters (See Heavy breed.)

Sitting Fertile eggs sold for incubation. Also a number of eggs that one hen can cover.

Soft feather Relatively soft plumage that 'lifts', found in chickens other than game birds.

Soya A source of protein in compound feeds.

Spangling Spots or splashes of different colours on the feathers.

Special Marketing Terms Descriptions adopted by the EU for labelling produce for sale. Systems must conform to requirements of stock density in relation to land use and housing. (See Free-range; Semi-intensive; Deep litter; Perchery/barn.)

Spice Vitamin and mineral supplement, traditionally given to help in re-feathering after the moult.

Split comb A division in the blade of a single comb.

Split crest The division of the crest or tuft of feathers on the head so that it falls on each side.

Split tail A gap in the base of the main tail feathers.

Split wing A space between the primary and secondary feathers of the wing.

Sport A random mutation which appears suddenly as a new characteristic, e.g. the White Wyandotte originated as a sport of the Silver Wyandotte.

Spur Horny growth on the shank of the male, used in fighting.

Squirrel-tail The tail carried so high that the feathers slope towards the back. Regarded as a show fault.

Staggy An old term for table cockerels when the flesh becomes more strong tasting and adult feathers appear as stubs, making them difficult to pluck.

Standard Scale of marks drawn up for each pure breed as an ideal to follow.

Sternum (See Keel.)

Strain A number of birds from the same family group of one breed.

Striping Dark line down the centre of a white feather, as in the neck hackles of Light Sussex and Columbian Wyandotte.

Stubs Sheaths of small quill feathers left behind on plucking.

Supplements Minerals and vitamins added to feeds.

Taint Unacceptable flavour, aroma or colour in meat or eggs as a result of coming into contact with an outside or inside agent.

Thighs Muscles covering area above the shank, always covered with feathers.

Thirds Middlings (See Wheat.)

Throat The upper and front area of the neck.

Tibia (See Drumstick.)

Ticking Small dark dots on feathers.

Tipping The tips of feathers are a different colour or shade from the rest.

Toe-punching Making a small hole in the web of the foot for identification purposes.

Trap-nest A nest that automatically closes behind the hen so that an egg can be identified as hers before she is released.

Treading Mating, as described by the action of the cock's feet on the back of the hen.

Tri-coloured A breed with the neck hackle, saddle hackle and wing bows of different colours.

Trio A group of one male and two females, often sold as a group by breeders of pure breeds.

Trussing Preparing a plucked and eviscerated bird for the table by securing the wings and legs with string or skewers.

Type The shape, size and conformation of a bird.

Uropygium (See Parson's nose.)

Utility chickens Those that have been bred for productive purposes, rather than for show.

Variety A sub-division of a breed, where colours and other features may vary, e.g., the Sussex breed has Light, White, Red, Brown and Speckled varieties.

Vaccination The introduction of vaccines into the system in order to promote antibodies and give immunity against specific viral diseases.

Vent Opening of the oviduct from which the egg emerges.

Vulture hock Stiff feathers growing outwards from the hock joint, as in the Sultan. It is a fault on most other breeds.

Walnut comb A descriptive name for a type of comb, as in the Brahma. Also called strawberry comb.

Waterglass Sodium silicate solution, traditionally used for storing eggs.

Wattles Fleshy lobes suspended from the jaw.

Weathering A yellow or brassy tint in white feathered birds such as the White Leghorn.

Web The skin between the toes or between the joints in the wings.

Wheat Feed grain. When milled, it produces four constituents – flour, middlings, sharps and fibre, in decreasing order of nutritive value.

Wing bay Triangular area of secondary wing feathers that is apparent when the wing is folded.

Wing clipping The annual practice of clipping the ends of the primary feathers on one wing to stop a flighty bird from escaping.

Wings The fore-limbs.

Wry tail A congenital deformity in which the tail is twisted to one side. Such birds should not become breeders.

Zoonoses Diseases that can be transmitted from poultry to humans.

BIBLIOGRAPHY

Specific references are detailed within the text or at the end of appropriate chapters.

Books

A Guide to Raising Chickens. Gail Damerow. (Storey Books, USA).
American Standard of Perfection, (American Poultry Association).
British Poultry Standards, Victoria Roberts, (Blackwells).
Diseases of Free-Range Poultry, Victoria Roberts (Whittet Books)
Important Poultry Diseases. (Intervet).
Incubation: A Guide to Hatching & Rearing (3rd ed), Katie Thear. (Broad Leys Publishing Ltd)
Modern Vermin Control. M & V Roberts (Domestic Fowl Trust).
Practical Chicken Keeping, Katie Thear (Ward Lock).
Organic Poultry Production. Nic Lampkin. (University of Wales).
Profitable Free Range Egg Production. Mick Dennett (Crowood).
Salsbury Manual of Poultry Diseases (Salsbury Laboratories).
The Chicken Health Handbook, Gail Damerow (Storey Books, USA).

Magazines

Country Smallholding – monthly publication for smallholders. 01392 444274. www.countrysmallholding.com
Smallholder – monthly publication for smallholders. www.smallholder.co.uk
Fancy Fowl – monthly publication for poultry fanciers. 01728 685832
Poultry World – monthly publication for poultry farmers. 01444 445566. www.reedbusiness.com

Videos

Poultry at Home, Victoria Roberts (Farming Press)
A Guide to Incubation on a Small Scale. (Fancy Fowl)

USEFUL ADDRESSES

Organisations

ADAS. 08457 7660085. www.adas.co.uk
British Egg Industry Council. 020 7370 7411. www.britegg.co.uk
British Free Range Egg Producers' Association. 01746 710817. www.bfrepa.co.net
DEFRA (Department of Environment, Food & Rural Affairs). Helpline: 08459 335577 Publications: 08459 556000. www.defra.gov.uk
Demeter/Bio-Dynamic Agriculture Association. 01453 759501

REMI (Regional Egg Marketing Inspectors).
Midlands & Wales: 01902 693145
North & North-East. 01132 309669
Eastern: 01223 462727
Southern: 01189 392215
Western: 01179 591000
FAWC (Farm Animal Welfare Council). 020 7904 6531
Freedom Food Ltd. 0870 444 3127. www.rspca.org.uk
Humane Slaughter Association. 01582 831919.

Irish Organic Farmers & Growers Association. 00 3531 830 7996
NFU (National Farmers' Union). 01572 824686. www.nfu.org.uk
Organic Helpline: 0117 922 7707
Organic Farmers & Growers Ltd. 01353 722398
Organic Food Federation. 01362 637314.
Poultry Club of Great Britain. 01205 724081. www.poultryclub.org
RSBT (Rare Breeds Survival Trust). 02476 696551. www.rare-breeds.com
Rare Poultry Society. 01162 593730
Scottish Organic Producers' Association. 01786 841657
Soil Association. 0117 914 2412. www.soilassociation.org
UKROFS (United Kingdom Register of Organic Food Standards). 020 7238 5915
Utility Poultry Breeders' Association. 01746 714 2109.

Courses

Easton College (Norwich). 01603 731200
Harper Adams University College. (Shropshire). 01952 820280
Scottish Agricultural Colleges. 0131 535 4000. www.sac.ac.uk

There are also a number of agricultural colleges, small farm training groups and other organisations offering short or part-time courses for small scale poultry keepers. They are normally listed in *Country Smallholding* magazine.

Training and Consultancy

ADAS. 01623 844331
Chris Belyavin (Technical) Ltd. 01952 813418
Mick Dennett. 01273 890592
Progressive Poultry Training. 01981 540308

Suppliers

Housing
Small houses (up to 100 birds)
Domestic Fowl Trust. 01386 833083. www.mywebpage.net/domestic-fowl-trust
Forsham Cottage Arks. 0800 163797.
www.forshamcottagearks.co.uk
Gardencraft. 01766 513036.
Hodgsons. 01833 650274.
Lifestyle UK Ltd. 01527 880078
Lindasgrove Arks. 01283 761510
Littleacre Products. 0121 308 2251.
The New Chislet Workshop. 01227 719133
Smiths Sectional Buildings Ltd. 0115 925 4722.
SPR. 01243 542815. www.sprcentre.co.uk

Large (including moveable) houses
ARM Buildings Ltd. 01889 575055. www.armbuildings.co.uk
Associated Poultry Services. 01884 257707
Fitzgerald Livestock Housing. 028 4273 8204. www.fitzhousing.com
Harlow Bros. Ltd. 01509 842561.
Liberty Livestock Systems. 01296 748842.
Morspan Ltd. 01291 672334. www.morspan.co.uk
NFP. 01531 631020. www.freerange.org
Patchett Engineering Ltd. 01274 882333. www.patchett.co.uk
Swiftranger. 01981 250076

Nest boxes
RJ Patchett Engineering Ltd. 01274 882333. www.patchett.co.uk
W Potter & Sons (Poultry) Ltd. 01455 553234.
Poultry Breeding Supplies. 01765 601810.

Lighting
ELM Ltd. 01933 223278.
Rooster-Booster Electronic System. 01963 34279. www.roosterbooster.co.uk

Electric Fencing & Netting
Bramley & Wellesley Ltd. 01452 300450. www.bramley.co.uk
Drivall Ltd. 0121 423 1122.
Electranets Ltd. 01452 617841.
Electric Fencing Direct. 01732 833976. www.electricfencing.co.uk
Hotline-Renco Ltd. 01626 331188. www.hotline-fencing.co.uk
Rappa Fencing. 01264 810665. www.rappa.co.uk
GA & MJ Strange. 01225 891236.

Feeding & Drinking Systems and Equipment
Autonest Ltd. 01536 760332.
BEC. 01623 662769. www.tatra.co.uk
Bird, Steven & Co. 01384 576381.
KJ Bower. 01845 578325.
EB Equipment Ltd. 01226 730037. www.eb-equipment.com
ELM. 01933 223278.
Hengrave Feeders Ltd. 01284 704803.
Lubing Aqua-2 Ltd. 01773 765316.
Newquip Ltd. 01765 641000.
Parkland Products. 01233 758650. www.parklandproducts.co.uk
Rainbow JFM. 01359 250238
Solway Feeders Ltd. 01557 500253. www.solwayfeeders.com

Litter
Arden Woodshavings. 01675 443888.
Dixons Dustless Ltd. (chopped straw). 01359 259341.
Environmental Straw Products. 01480 860745.
NS Milling Ltd. (chopped straw). 01405 860011.
Robin Foster-Clarke. (wood shavings, chopped straw). 01986 785278.
Snowflake Woodshavings Co. Ltd. 01205 311332.
Straw Services Ltd. 0113 282 6164.

Feeds
AF plc. 01772 799330.
Allen & Page. 01362 822900. www.smallholderfeed.co.uk
J Bibby Agriculture Ltd. 01733 555552.
BOCM Pauls Ltd. 01473 232222. www.bocmpauls.co.uk
Clark & Butcher Ltd. 01353 720237.
Crediton Milling Co. Ltd. 01363 772212.
HI Glasser. 01442 890100.
Humphreys Farms. 01962 764523.
Lloyds Animal Feeds Ltd. 01691 830741.
W & H Marriage & Sons Ltd. 01245 612000. www.marriagefeeds.co.uk
Vitrition. 01777 228741. www.optivite.co.uk
Yorkshire Country Feeds. 01609 780140.

Egg Packaging & Labelling
Clear-Sell Ltd. 01438 742400.
Conexpak. 02476 221990.
Danro Ltd. 01455 847061. www.danroltd.co.uk
Omni-Pac UK. 01493 855381.
Van Leer. 028 3832 7711.

Egg Packers/Distributors
Deans Foods Ltd. 01442 891811.
Freshlay Foods Ltd. 01409 261242. www.freshlay.co.uk
Stonegate Farmers Ltd. 01323 846565.
The Lakes Free Range Egg Co. Ltd. 01768 890460.

Incubators
Aliwal Cabinet Incubators. 01508 489328.
Brinsea Products Ltd. 01934 823039. www.brinsea.co.uk
Curfew Incubators. 01621 741923. www.curfew.co.uk
Interhatch. 0700 4628 228.
MS Incubators. 0116 247 8335
Natureform Hatchery Systems. 01608 686591. www.natureform.com
Southern Aviaries. 01825 830930.

Health and Hygiene Products & Services
Agil Ltd. 0118 981 3333. www.agil.com
Agquip Ltd. 01242 621258.
Ainsworths. (homoeopathic products). 01271 342077.
Anglian Farm Supplies. 01842 765634.
Animal Aids Ltd. 01963 33083
Antec International Ltd. 01787 377305. www.antecint.com
Barrier Animal Healthcare. 01953 456363.
Bayer plc. 01284 763200.
Breckland International Ltd. 01760 756414.
Climatec Systems. 01531 631161.
Diversey Lever Ltd. 01604 783505.
Intervet UK Ltd. 01223 420221.
Janssen Animal Health. (Flubenvet) 01494 567555.
Lohmann Animal Health. 01489 576093
Mike Bowden Livestock Services. 01953 851799
Merial Animal Health. 01279 775858.
RS Hygiene. 01638 71444

Schering-Plough Animal Health. 01895 626000.
Sci-Tech Laboratories. 01588 672600.
Technical Services Consultants. 01706 620600
The Hydor Co. Ltd. 01725 511422.
Uff Hygiene. 01507 602707.

Veterinary Services
Crowshall Veterinary Services. (Norfolk) 01953
    455454.
DG Parsons. (Wilts) 01225 790090
David Spackman. (Somerset) 01278 661007
Minster Veterinary Practice. (North Yorks)
    01904 643997.
Retford Poultry Partnership. (Notts) 01777
    703011.
Sandhill Veterinary Services. (North Yorks)
    01845 578710.
Sci-Tech Laboratories. (Shropshire) 01588
    672600.
Slate Hall Veterinary Practice. (Cambs) 01954
    789424.

Vermin Control Products
Acorn Pest Control. 02476 491689.
Amtex Ltd. 01568 610900.
    www.thezapper.co.uk
Phoenix Products. 01584 711701.
Sorex Ltd. 0151 420 7151. www.sorex.com

Security Products
JCS Electrical. 01530 263263.
Securcom Systems Ltd. 07041 544142
Sutcliffe Electronics. 01233 634191.
    www.sutcliffe-electronics.co.uk

Insurance
Greenlands Insurance. 01970 615561.
    www.greenlands.co.uk
Hodgson Barrow Ltd. 0151 227 3731.
NFU Mutual. 01789 202411.
Ottery Insurance Services. 01404 813495.
    www.ottery.co.uk

Computer Software
FOL Software. 0116 202 2000.
    www.farming.co.uk
Chris Belyavin (Technical). 01952 813418.

www.flockdata.com
QuickBooks. 0800 374285. www.bsd-
    online.co.uk

Miscellaneous Equipment & Supplies
Ascott Smallholding Supplies. 0845 130 6285.
    www.ascott-shop.com
AXT-Electronic (Electronic pop-hole control-
    ler). 0049 369 23 50 424 (Germany).
    www.axt-electronic.de
Barry Marlin. (cleaning). 01825 840660.
Beefi Startin. (wing tags). 01750 22940
Cyril Bason (Stokesay) Ltd. 01588 673 0204.
E Collinson & Co. Ltd. (silos). 01995 606451
EB Equipment Ltd. (silos). 01226 730037
ELM. 01933 223278.
Gamlins Farm Ltd. (cleaning). 01823 672596
JB Services. (cleaning). 01449 673232.
Onduline Building Products Ltd. (roofing). 020
    7727 0533.
One Stop Poultry Shop. 01379 586288.
    www.onestoppoultryshop.com
Oxmoor Smallholder Supplies. 01757 288186.
PCG Ltd. (egg weighing scales). 01724 734025.
Peter Pell Poultry Services. (cleaning). 01205
    280391.
Reids Used Equipment Sales. 01789 720027.
    www.reidgroup.co.uk
Roxan ID. (leg rings). 01750 76226.
Stock Nutrition. (plucking machines). 01362
    851200.
The Domestic Fowl Trust. 01386 833083.
    www.mywebpage.net/domestic-fowl-trust
Whitehead Engineering. (plucking machines).
    01761 432305.
Woodhurst Garden Fowl. 01487 822053.
Woodside Poultry & Livestock Centre. 01582
    841044. www.woodsidefarm.co.uk

Stock
Traditional breeds are available from poultry
breeders throughout the country. Consult the
Poultry Club of Great Britain for details of
breed clubs. Breeders are listed in the
Breeders' Directory which is updated
monthly in Country Smallholding magazine
and on their website:

www.countrysmallholding.com

Collections of Traditional Breeds on View

The *Domestic Fowl Trust*, Honeybourne, Evesham, Worcestershire WR11 5QG. 01386 833083.

South of England Rare Breed Centre, Woodchurch, Ashford, Kent TN26 3RJ. 01233 861493.

The Wernlas Collection, Onibury, Ludlow, Shropshire SY7 9BL. 01584 856318.

Suppliers of free-range hybrids to small-scale poultry keepers

Cyril Bason (Stokesay) Ltd. (layers and table birds). 01588 6730204

Meadowsweet Poultry Services (layers). 0191 384 2259. www.meadowsweetpoultry.co.uk

Muirfield Hatchery. (layers). 01577 840401. www.theblackrock.co.uk

Piggotts Poultry Breeders. (layers). 01525 220944.

S&T Poultry. (table birds). 01945 585618.

Hatcheries and Pullet Rearers

AJ Hood Growing Pullets. 01794 515438.

Avitec Ltd. (Scotland). 01307 465906.

B & B Enterprises. 01790 763066.

Blue Barns Hatcheries Ltd. 0191 4102809.

Burcombe Farm (organic chicks). 01769 550330. www.traditional-organics.co.uk

Country Fresh Pullets Ltd. 01691 831020.

Farm-Fresh Hatcheries Ltd. 01772 814081.

Grassington Rangers Ltd. 01825 723253.

Hubbard-ISA Ltd. 01733 223333.

Hy-Line UK Ltd. 01564 703704.

Johnston's of Mountnorris Ltd. 01861 507281. www.johnstons-of mountnorrisfreeserve.co.uk

Joice & Hill Poultry Ltd. 01328 838216.

Lohmann GB Ltd. 01526 351009.

Maple Leaf Chicks. 01772 824534.

MJ Hayward & Sons. 01425 652007.

Maurice Millard (Hatcheries) Ltd. 01225 722215.

Mayfield Chicks Ltd. 01706 229641.

PD Hook. 01993 850261.

Potters Poultry. 01455 553234.

Poultry First Ltd. 01526 352471.

RA Wright (Chicks) Ltd. (Northern Ireland). 02892 664231.

Ross Breeders Ltd. 0131 3331056. www.rossbreeders.com

Ross Poultry Ltd. 01526 352471.

Shaver Poultry Breeding Farms. 01772 824534.

SJ & DR Blake. 01566 775366.

The Cobb Breeding Co. Ltd. 01245 400109. www.cobb-vantress.com

Tom Barron Ltd. 01772 690111. www.tombarron.co.uk

Whitakers Hatcheries Ltd. (Ireland). 00353 (0) 21 374466.

USA sources

It is necessary to check whether there are any zoning regulations which may prohibit the keeping of poultry in certain areas. The US Department of Agriculture (USDA) will also provide information that applies on a national basis. It is necessary, for example, to register with them if poultry is processed for sale. Some regulations vary between states, but information is available from state, county and city agencies. These are listed in local telephone directories. The best starting point, however, is to contact the Extension Poultry Specialist of the appropriate state university or state department of agriculture:

Alabama 205 844 2613.
Alaska. 907 474 6357.
Arizona. 602 621 1980.
Arkansas. 501 543 8526.
California. 916 752 3513.

Louisiana. 504 388 4481
Maine. 207 581 2768
Maryland. 410 651 9111.
Massachusetts. 413 545 2432.
Michigan. 517 353 2906.

Ohio. 614 292 4821.
Oklahoma. 405 744 9293.
Oregon. 503 737 2254.
Pennsylvania. 717 394 6851.
Puerto Rico. 809 832 4040.

Colorado. 303 491 7803.
Connecticut. 203 486 1008.
Delaware. 302 856 7303.
Florida. 904 392 1931.
Georgia. 706 542 1325.
Hawaii. 808 956 8334.
Idaho. 208 459 6365.
Illinois. 217 244 0195.
Indiana. 317 494 8009.
Iowa. 515 294 4303.
Kansas. 913 532 6533.
Kentucky. 606 257 7529.

Minnesota. 612 624 4928
Mississippi. 601 325 3416.
Missouri. 314 882 6754.
Montana. 406 293 7781.
Nebraska. 402 472 6451.
Nevada. 702 397 2184.
New Hampshire. 603 862 2247.
New Jersey. 908 932 9702.
New Mexico. 505 646 3016.
New York. 607 255 8143.
North Carolina. 919 515 5391.
North Dakota. 701 237 7691.

Rhode Island. 401 792 2072..
South Carolina. 803 656 4026.
South Dakota. 605 693 3484.
Tennessee. 615 974 7351.
Texas. 409 845 4318.
Utah. 801 750 2162.
Vermont. 802 656 2074.
Virginia. 703 231 5087.
Washington. 206 840 4579.
West Virginia. 304 293 2406.
Wisconsin. 608 262 9764.
Wyoming. 307 766 3100.

Organisations

Agricultural Network Information.
   www.agnic.org
American Bantam Association. 973-383-6944
American Farm Bureau Federation. www.fb.org
American Food Safety Standards.
   www.foodsafety.gov
American Livestock Breeds Conservancy. 919 542
   5704. www.albc-usa.org
American Poultry Association. 508-473-8769
American Pastured Poultry Producers' Associa-
   tion. 715-723-2293
Appropriate Technology Transfer for Rural Areas.
   www.attra.org
Kerr Center for Sustainable Agriculture.
   www.kerrcenter.com
Organic Farmers' Marketing Association.
   www.web.iquest.net
Produce Marketing Association.
   www.aboutproduce.com
Small Farm Resources.
   www.farminfo.org
Society for the Preservation of Poultry Antiqui-
   ties. 319 246 2299
U.S. Department of Agriculture (USDA).
   www.usda.gov
USDA Agricultural Marketing (farmers'
   markets). 1-800-384-8704

Magazines

Countryside & Small Stock Journal. 1 800 551
   5691
Poultry Press. 765 827 0932
Small Farm Digest. www.reeusda.gov/
   smallfarm

Small Farm Today magazine.
   www.smallfarmtoday.com

Suppliers

AB Incubators Ltd.
(small incubators). 309 793 4273.
BF Products Inc. (fencing and netting). 717 238
   7715.
Brower. (pasture pens and feed equipment).
   www.browerequip.com
Common Sense Fence. (fencing)
   www.geoteking.com
Endurance Net. (all types of netting). 1-800-
   808-6387.
Heritage Building Systems. (steel farm build-
   ings). 800-643-5555.
Humidaire Incubator Co. (incubators). 937 996
   3001
Jeffers Vet Supply. 800 533 3377.
Lyon Electric Co. (incubators and brooders).
   619 216 3400
Murray MacMurray Hatchery. (traditional and
   rare breeds).1-800-456-3280.
National Band & Tag. 859 261 2035
Orscheleln Farm & Home. (farm equipment).
   www.orschelnfarmhome.com
Patterson Poultry Supplies. 540 638 2297
Port-A-Hut (portable ark shelters).
   www.port-a-hut.com
Purina Mills. (feeds). www.purinamills.com
Stromberg's Chicks. 218 587 2222.
Tomahawk Live Trap Co. 715 453 3500.

Index

Numbers in italics indicate tables, figures or illustrations

other books from Whittet Books

that you might like to buy

POULTRY FOR ANYONE by Victoria Roberts £19.99 large format hardback
Full colour guide to the breeds of chicken with description of their history, characteristics, special requirements and utility: 90 colour photographs. Plus information on management and showing.

DUCKS, GEESE AND TURKEYS for Anyone by Victoria Roberts £19.99 large format hardback
Full colour guide to the breeds of ducks, geese and turkeys with description of their history, characteristics, special requirements and utility, plus information on management and housing. 70 colour photographs.

DISEASES OF FREE RANGE POULTRY by Victoria Roberts £15.99 hardback
Comprehensive guide in colour to the ailments that your poultry may be suffering from, and what to do about it. Covers other domesticated bird species, such as ducks, geese, turkeys, guinea fowl, quail and pheasants. Diagnosis chart. Fully illustrated (including colour)

ALL ABOUT GOATS Lois Hetherington and John G Matthews £15.99 large format hardback
Everything you need to know about care, management, choosing, showing and also goat products. Fully illustrated (including colour)

LOOKING AFTER A DONKEY Dorothy Morris £10.99 paperback
How to choose your donkey and then how to look after it, including feeding, housing, driving, riding, breeding and showing. Many photographs illustrate text.

Please send cheque (including £1.50 for postage and packing) to

Whittet Books Ltd
Hill Farm
Stonham Rd
Cotton
Stowmarket
Suffolk IP14 4RQ

If you would like to be added to our mailing list for catalogues, please write to the same address.